Synthesis Lectures on Electrical Engineering

This series of short books covers a broad spectrum of titles of interest in electrical engineering that may not specifically fit within another series. Books will focus on fundamentals, methods, and advances of interest to electrical and electronic engineers.

Blas M. Vinagre

Time in Control Theory

On Concepts, Measures and Uses

 Springer

Blas M. Vinagre
Department of Electrical, Electronic
and Automation Engineering
University of Extremadura
Badajoz, Spain

ISSN 1559-811X   ISSN 1559-8128 (electronic)
Synthesis Lectures on Electrical Engineering
ISBN 978-3-031-54041-7   ISBN 978-3-031-54042-4 (eBook)
https://doi.org/10.1007/978-3-031-54042-4

This Springer imprint is published by the registered company Springer Nature Switzerland AG
The registered company address is: Gewerbestrasse 11, 6330 Cham, Switzerland

Paper in this product is recyclable.

That the idea of time, ..., is the result of a long evolution, it is difficult to deny.

Jean-Marie Guyau. *La genèse de l'idée de temps.*

Time, so Austerlitz said at the Greenwich observatory, was by far the most artificial of all our inventions...

W. G. Sebald, *Austerlitz.*

Soon it became evident that the concept of time is curiously evasive when looked at from the separate and necessarily limited viewpoints of individual disciplines. For example, the physicist symbol t is a deceivingly simple representation of «what we mean by time.»

J. T. Fraser, *The Voices of Time.*

Preface

> *But at my back I alwaies hear*
> *Time's winged chariot hurrying near:*
> Andrew Marvell, *«To his Coy Mistress»*

The late American writer David Foster Wallace, best known for his almost infinite novel *Infinite Jest*, invited to give a lecture to graduates of Kenyon College (OH), began with the following story [1]:

> There are these two young fish swimming along and they happen to meet an older fish swimming to the other way, who nods at them and says "Morning, boys. How's the water" And the two young fish swim on for a bit, and then eventually one of them looks over at the other and goes "What the hell is water?"

For us, as humans and as control engineers, time is the water in which we swim. But «what the hell is time?» With Augustine, Bishop of Hippo at the end of the 4th century and the beginning of the 5th, we can say [2, XI, XIV, 17]:

> What, then, is time? If no one asks me, I know; but if I want to explain it to the person who asks me, I don't know.

And we don't know, too, even seventeen centuries later and after huge efforts for conceiving, measuring and using it. Peter Brueghel's *The Triumph of Time* (see Fig. 1.1), is a panoply of symbols and ideas used along the history when dealing with time. An old man is devouring a little child that he holds with his right hand, while with his left hand, he lifts a snake that, forming a circle, bites its tail. The old man is over a moving chariot sitting on an hourglass, and above him, hanging from a tree rooted in a sphere surrounded by the zodiac signs, there is a weight driven mechanical clock. On the right side, in front of the chariot, all things, from the sky and from the ground, are images of spring. On the left side, behind the chariot, with the background of a storm, a monk dressed skeleton who rides a donkey and carries a scythe is followed by a cherub riding an elephant and blowing a trumpet. As chariot advances carrying the old man and the tree, the mandala type wheels of the chariot crush the objects spread on the ground, everyday objects from home, industry, arts, or science.

Of course, after the scientific revolution, Father Time changed some of his dresses and symbols, and nowadays, at the doors of a new revolution, he carries a heavy burden of hypothesis and chains. So, I'll not dare here to try to explain him. My purpose is much simpler: following a quasi-chronological order, to expose, I hope in a not very tedious way, how the concepts of time and the ways of measuring and using it are reflected in some of the developments of the *Control Theory* from its origins to the present day. In what follows, we will explore issues such as the advent and development of pendulum clocks and their contributions to automatic control, the search for the minimum time and the origins of optimal control, the birth of the control theory with the calculus and the Newtonian concept of time, the relations between sampling, integral sums and the relational concept of time defended by Leibniz, or the concept of time reversibility and its influence on dynamical systems. Finally, we will give some insights about the role of time in PID control, the relation of feedback with memory and time, and some considerations on time for Artificial Intelligence.

Given the «unity and diversity of time»,[1] a subject of philosophy and religion, psychology and life sciences, as well as social and physical sciences, the approach used in this work, though brief and focused because it deals with an engineering discipline, tries to be panoramic. Given the disequilibrium between the enormity of the subject (time) and the size of the work, some readers could consider this little book as a case of «Parturient montes, nascetur ridiculus mus».[2] It is my sincere hope that for others be at least an interesting *divertimento*. In any case, the result is due to both, my speculation and my ignorance.

Badajoz, Spain
December 2023

Blas M. Vinagre

---

[1] These words, as well as many interesting ideas, are borrowed from the wonderful collective book edited J. T. Fraser, *The Voices of Time* [3].

[2] «Mountains will go into labour, and a silly little mouse will be born» (Horace, *The Art of Poetry*, 139).

References

1. David Foster Wallace. Kenyon commencement address. https://web.ics.purdue.edu/drkelly/DFW KenyonAddress2005.pdf, (2005)
2. San Agustín. *Confesiones*. Bruguera, 1984
3. J. T. Fraser. *The Voices of Time*. George Braziller, 1966

Acknowledgements

The original work presented here is part of a line of research on the relationships and common bases of different areas of human knowledge: technology, science, medicine, and humanities. This line of research responds to personal and professional interests that the author, although without public results until very recently, has maintained over many years. The opportunity to put these results in writing arose in 2015 from the invitation to give a plenary lecture at the XI Symposium on Intelligent Control of the Comité Español de Automática. I would like to thank the Organizing Committee of that symposium, and especially its president, Dr. Antonio José Calderón Godoy, for his kind invitation. From the labour to prepare said conference, the work presented here emerged.

Likewise, my sincere and friendly thanks to Prof. Igor Podlubny, at the Technical University of Kosice, for many good times of conversation on time and other topics.

Finally, I would like to thank Springer Nature in the persons of Dieter Merkle and Ramasubramaniyan Velu, both the interest shown in receiving the manuscript, and the work and patience required until we had a final version.

Concerning the contents and figures, acknowledgement is given to the respective copyright holders to reproduce material from the following sources:

Figure 1.1: Public Domain, https://commons.wikimedia.org/w/index.php?curid=816 55749.

Figure 1.2: Eastern and Egyptian scenery, ruins, &c. : accompanied with descriptive notes, maps, and plans, illustrative of a journey from India to Europe : followed by an outline of an overland route, statistical remarks, &c. intended to shew the advantage and practicability of steam navigation from England to India by Charles Franklin Head. Online version by the Getty Research Institute; the Cleopatra's Needle image is on page 106 of 150: https://archive.org/.

Figure 1.3: Public Domain, https://commons.wikimedia.org/w/index.php?curid=323 68149.

Figure 1.4: Public Domain, https://commons.wikimedia.org/w/index.php?curid=202 8026.

Figure 1.5: Public Domain, https://commons.wikimedia.org/w/index.php?curid=211 4036.

Figure 1.6: Public Domain, https://commons.wikimedia.org/w/index.php?curid=330 69391.

Figure 1.7: Public Domain, https://commons.wikimedia.org/w/index.php?curid=287 6455.

Figure 1.8: Public Domain, https://commons.wikimedia.org/w/index.php?curid=288 9696.

Figure 1.9: Public Domain, https://commons.wikimedia.org/w/index.php?curid=967 1892.

Figure 2.4: Public Domain, https://commons.wikimedia.org/w/index.php?curid=544 6377.

Figure 2.6: By Mikerollem—Own work, CC BY-SA 4.0, https://commons.wikimedia.org/w/index.php?curid=114606580.

Figure 6.8: Public Domain, https://commons.wikimedia.org/w/index.php?curid= 198284.

Figure 6.9: Public Domain, https://en.m.wikipedia.org/wiki/.

Section 6.1: some material reproduced from reference [1].

Section A.3: some material reproduced from references [2, 3].

References

1. Inés Tejado, Blas M. Vinagre, José Emilio Traver, Javier Prieto-Arranz, and Cristina Nuevo-Gallardo. Back to basics: Meaning of the parameters of fractional order pid controllers. *Mathematics*, 7(530):2 of 16, 2019

2. D. Sierociuk and B. M. Vinagre. State and output feedback fractional control by system augmentation. In *Proceedings of the 4th IFAC Workshop Fractional Differentiation and its Applications (FDA2010)*, Badajoz, Spain, 2010

3. Inés Tejado, Blas M. Vinagre, and Dominik Sierociuk. State space methods for fractional controllers design. In Ivo Petráš, editor, *Handbook of Fractional Calculus with Applications. Volume 6 Applications in Control*, pages 149–174. De Gruyter, 2019

Contents

List of Figures

The quintessential symbol for authority, of course, was the clock.

Otto Mayr, *Authority, Liberty & Automatic Machinery in Early Modern Europe.*

1.1 A Moving Image of Eternity

Plato's theory of knowledge is summarized in one of his most famous myths, that of the cave [1, VII, 514–517]. In it, he proposes that we imagine some captives who, chained in such a way that they cannot move, have lived since childhood in an underground cave whose frontier wall is their only world. Behind them, objects and beings pass by along a steep path, and still further back a fire illuminates them and projects on the wall the silhouettes of what is passing between the fire and them. If we freed these captives from their chains, they would have the opportunity to abandon their ignorance that makes them consider the shadows on the wall as the only reality, since turning around they could contemplate the fire that makes them possible and the realities of which they are a projection. Even more, the one who managed to get out of the cave could contemplate the ultimate «cause of what he saw in the cave with his fellow captives». This ascent from darkness to light, from the sensible world to the intelligible world, from appearances to forms, is also that of the transitory to the eternal, from the temporal to the timeless. Plato's theory of knowledge is a mirror image of his theory of the creation of time as expounded in the Timaeus [2, 37 d]:

> The nature of the Living Being was eternal, and it was not possible to bestow this attribute fully on the created universe; but he determined to make a moving image of eternity, and so when he ordered the heavens he made in that which we call time an eternal moving [according to number] image of the eternity which remains for ever at one.

Inserted into nature, unconscious like animals and children, our ancestors limited themselves to suffer this «eternal moving image of the eternity.» Curious, they later learned to

B. M. Vinagre, *Time in Control Theory*, Synthesis Lectures on Electrical Engineering, https://doi.org/10.1007/978-3-031-54042-4_1

Fig. 1.1 Peter Breughel, *The triumph of time*

observe it. Later on, "Out of Plato's cave" [3], they learned to measure it, to manage it and, finally, to create it. But it is possible that in each of these stages we have been dealing with different representations of time, different invocations, and that the concepts and attributes with which it has reached the current era, be just a result of accretions and metamorphoses throughout the human history, as has happened with Father Time (Fig. 1.1) in Western iconography: perhaps originating in a confusion between the Greek term and ancient God for time, Chronos, and the name of the darkest of their Titans, Kronos, among their attributes we can find not only symbols of transience, creative eternity or cosmic powers, the wings, the zodiac or armillary spheres, but also of ruin, consumption and destruction, an elderly man with the hourglass and the scythe [4].

1.2   Sundials and Water Clocks

The oldest timekeeper was the sundial, which provides precision to the simple observation of the everyday sunrise and sunset, as well as of seasonal ascension and declination. Originating in Babylon or Egypt, their trace can be followed through the Fertile Crescent, and the arriving to Greek and Roman cultures is present in the testimony of their writers and historians. Thus Herodotus tells us [5, II, 109, 3]: «... the Greeks learned the pole, the gnomon, and the division of the day into twelve parts from the Babylonians.»

Probably, humans first learned to estimate the hour by measuring the shadow of a mountain, of a tree or of their own bodies. So, mountain, tree or body were used as gnomons: a vertical structure indicating the time by the length of its projected shadow, as the obelisk known as *Cleopatra's Needle* (Fig. 1.2), installed in Alexandria. Another one was brought to the Roman forum and led a character from Plautus, a wanderer, to exclaim [6, III, III]: «Gods damn that man who invented the hours, and installed in this city the first quadrant!»

Fig. 1.2 Cleopatra's Needle in old Alexandria

Fig. 1.3 Horizontal sundial located at flagstaff gardens, Melbourne, Australia. The gnomon is the triangular blade. The style is its inclined edge

Later, people learned to match the measured shadow according to the season of the year by drawing concentric circles around it, the angle of inclination of the gnomon or stilus to the geographical latitude, and they do all this so that the estimate of the time would be reasonably accurate in any season of the year and in any place on Earth. It was the beginning of a new subject of study, the *gnomonics* or the science of sundials, which combines geometry and astronomy [7]. The simplest sundial consists of a flat plate (the dial) and a gnomon, which projected a shadow onto the dial. As the Sun appears to move through the sky, the shadow aligns with different hour-lines, which are marked on the dial to indicate the time of day. The style is the time-telling edge of the gnomon, though a single point or nodus may be used. The gnomon casts a broad shadow; the shadow of the style shows the time. The gnomon may be a rod, wire, or elaborately decorated metal casting. The style must be parallel to the axis of the Earth's rotation for the sundial to be accurate throughout the year. The style's angle from horizontal is equal to the sundial's geographical latitude (Fig. 1.3). With the sundial, humans begin to separate day and night, and, which is more important, to divide them into twelve parts with seasonal varying lengths, the so called *temporal or seasonal hours*: since summer days are longer than winter days, the corresponding summer hours were longer than winter hours [8].

Fig. 1.4 An early 19th-century
illustration of Ctesibius's
(285–222 BC) clepsydra from
the 3rd century BC

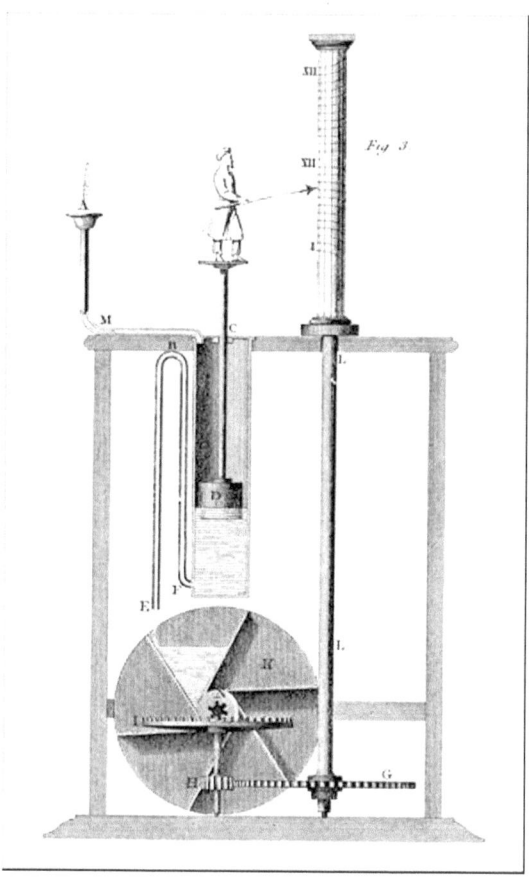

But the sundial give no information if it is cloudy, and provides no other output. For solving these and other drawbacks, like low precision, a step forward was done with the invention of the water clock or the *clepsydra* (from the Ancient Greek 'to steal water'). The essential elements of the device are a water vessel or container with a slowly changing level that, tracked by a float and transmitted through levers, wheels, gears and linkages, served not only as an element for measuring time, but also as a source or power that could drive sophisticated and large mechanisms. In Fig. 1.4 we can see an early 19th-century illustration of Ctesibius's (285–222 BC) clepsydra from the 3rd century BC: water that fall into the vessel operates a series of gears that rotates a cylinder with marked hours of varying lengths (the temporal hours) appropriate for the date; the hour indicator ascends as water flows in, and the overall design, including the small opening that allows water to enter and the rest of the mechanisms, make the rate of flow uniform in order to afford an enough accurate measure of time [9].

So, in addition to a greater precision in measuring the time and a total independence of external phenomena, the water clock made two important contributions to our history: it was an *automaton* and a *self-adjusting machine*: «it taught how to regulate machinery over long spans of time by means of program control» [8, p. 4]. And it paved the way for the mechanical clock.

1.3   The First Modern Machine

From the beginning, as we have seen, the measurement of time was attempted through a physical phenomenon that took advantage of or emulated some characteristics of time observed in external nature: continuity, uniformity and unidirectionality. Thus, sundials were succeeded by water clocks and hourglasses, making either the water or the sand flow continuously, just as time apparently flows. The invention of the mechanical clock meant oblivion of the cosmic and natural rhythms and the adoption of rigid units to measure time, the change from temporal hours to equinoctial ones [8]. The machine age began when the idea of naively and humbly following nature was abandoned in favor of replicating or replacing it; and use counting instead of observing a course to make the measurement [10, 11].

The mechanical clock has a tangled history, some people arguing that, anticipated by the automated gadgets of the antiquity, including water-clocks, it had a long tradition in Chinese culture before arriving to Europe [12, 13]. In any case, the device was first driven by weights or springs, and the periodic motion obtained by wheels and controlled by the *escapement*. This mechanism was conceived to regulate the rate of the clock by allowing the gear train to advance at regular intervals. In Figs. 1.5 and 1.6 are shown two early kinds of escapements: Verge and foliot and Verge with balance wheel. The foliot was a horizontal bar with weights near its ends affixed to a vertical bar called the verge which was suspended free to rotate.

Fig. 1.5 Verge-and-Folot escapement

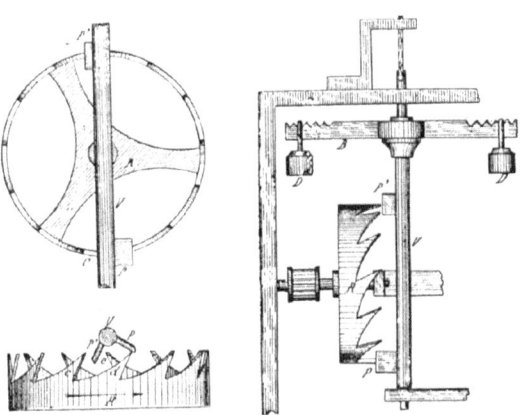

Fig. 1.6 Verge escapement and
balance wheel

The verge escapement caused the foliot to oscillate back and forth about its vertical axis. The rate of the clock could be adjusted by moving the weights in or out on the foliot. It was a kind of primitive balance wheel, a weighted wheel that rotates back and forth, being returned toward its center position by a spiral torsion spring. As the foliot, it is driven by the escapement, which transforms the rotating motion of the watch gear train into impulses delivered to the balance wheel.

On the other hand, spring powered clocks supposed and advance over the weight driven ones because they were compact and capable of storing energy for long periods of operation, these two characteristics meaning portability and the possibility of domestic and personal uses, as well as its integration as parts of more complex machines. In Fig. 1.7 is shown one of the first solutions, a *barrel and fusee* spring mechanism. The fusee is the cone-shaped pulley with a helical groove around it, wound with a cord or chain attached to the mainspring barrel. The torque provided by a mainspring decreases linearly as the spring unwinds during a clock's running period. The fusee's purpose is to even out this torque. In Fig. 1.8 the device is incorporated into a machine [9].

But the most perfect realization of the mechanical clock came with the pendulum clock in the 17th century. Galileo already worked on it as a part of his studies about falling bodies (supposing the body moving down a succession of small inclined planes), and was aware of its application as a mechanism for improving the precision of clocks, having the idea of using a swinging bob to regulate the motion of these devices (around 1640). But the definitive contribution was made by Christiaan Huygens in his analysis of circular motion as a combination of centrifugal and centripetal forces. The pendulum was approached as a particular case, and his results in the form of generalized formulas serve to inspire both, practical applications and further studies. In his work *Horologium oscillatorium* (1673) he

Fig. 1.7 Fusee and mainspring, showing operation. The fusee (cone-shaped pulley) was used to improve accuracy by equalizing the force of the mainspring

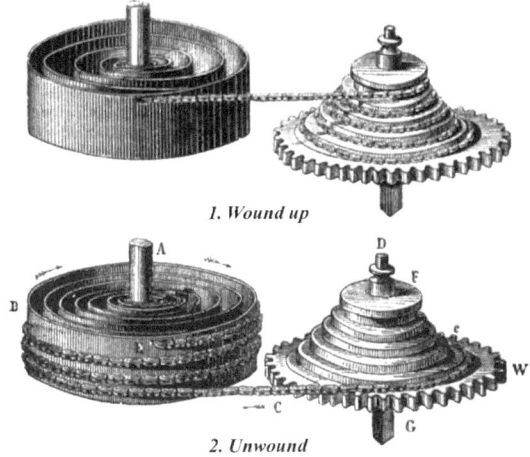

1. Wound up

2. Unwound

Fig. 1.8 Drawing of a machine incorporating a fusee, by Leonardo da Vinci circa 1490

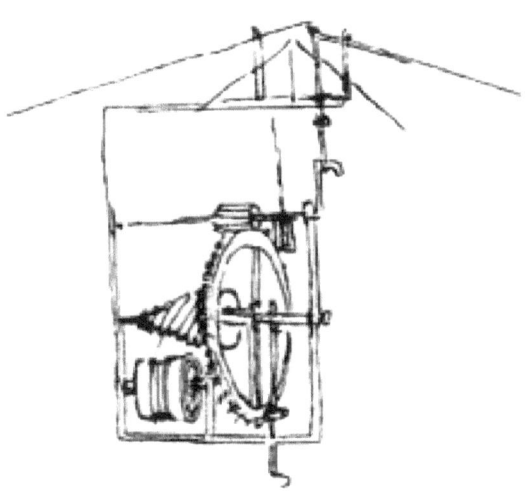

gave a complete theory of the machine (Fig. 1.9), and concluded that the beat of a pendulum is really isochronous only when the pendulum swings following a cycloidal arc[1] [9].

Subsequently, many refinements were made, the majority of them affecting the escapement mechanism or in order to reduce the size, by scientist and clockmakers, but the essential of the design was preserved [7–9, 14]. Figure 1.10 shows some milestones in clocks accuracy [15].

By using nowadays terminology, the pendulum can be figured out as the two blocks feedback system in Fig. 1.11, one representing the pendulum and the other one the escape-

---

[1] We will say more about this issue when studying optimal control problems in Chap. 2.

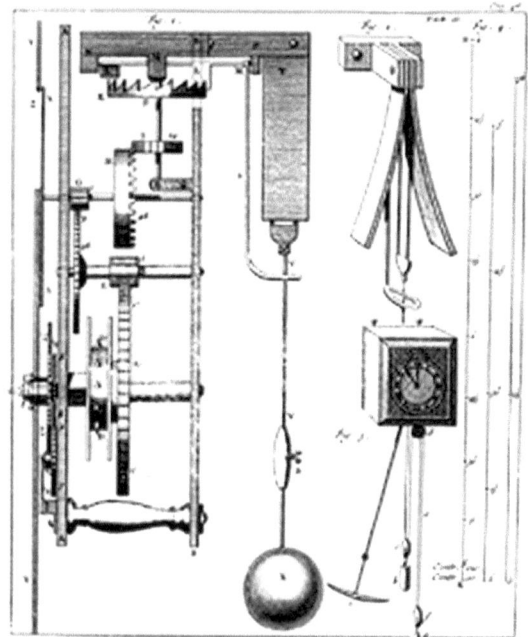

Fig. 1.9 Huygens pendulum clock

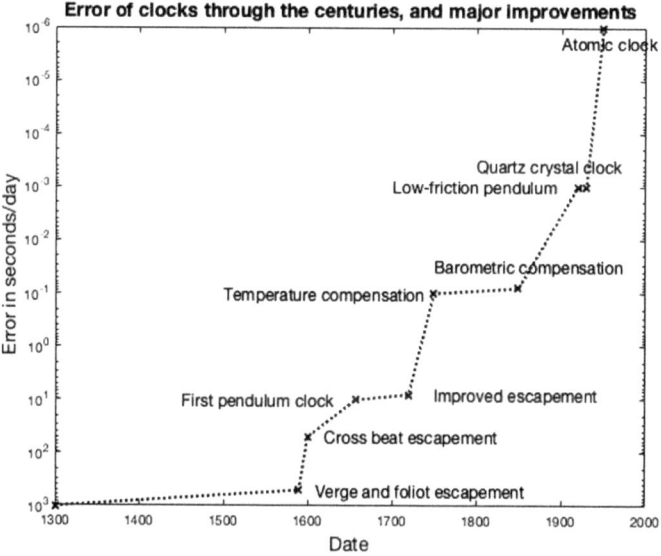

Fig. 1.10 Major milestones in clock accuracy (adapted from [16])

Fig. 1.11 Block diagram of the operation of a pendulum clock with escapement mechanism

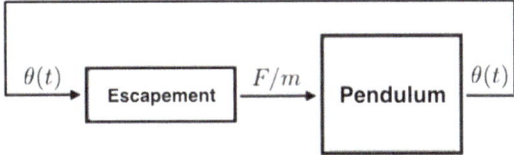

ment, where $\theta(t)$ is the angle of the pendulum, m the mass and F the force exerted by the escapement mechanism.

The equation of motion can be formulated as:

$$\frac{d^2\theta(t)}{dt^2} = -\frac{c}{ml^2}\frac{d\theta(t)}{dt} - \frac{g}{l}\mathrm{sen}(\theta(t)) - \frac{F}{m}, \tag{1.1}$$

being l the length, c the coefficient of friction, and g the acceleration of gravity. The operation of the escapement mechanism can be formulated, in a general way, as follows:

$$F/m = \begin{bmatrix} K\,\mathrm{sign}(\theta(t)), & \text{if } |\theta_m| \le |\theta(t)| \le |\theta_M| \; \frac{d|\theta(t)|}{dt} < 0 \\ 0, & \text{other} \end{bmatrix} \tag{1.2}$$

being $\mathrm{sign}(\cdot)$ the sign function, and $[|\theta_m|, |\theta_M|]$ the interval of angles where the escapement actuates. A complete block diagram ready for simulation using MATLAB$^\copyright$ and Simulink$^\copyright$ software is presented in Fig. 1.12 [17]. The detail of the escapement block is given in Fig. 1.13, and the outputs of both, the angle of the pendulum and the force exerted by the escapement are presented in Fig. 1.14.

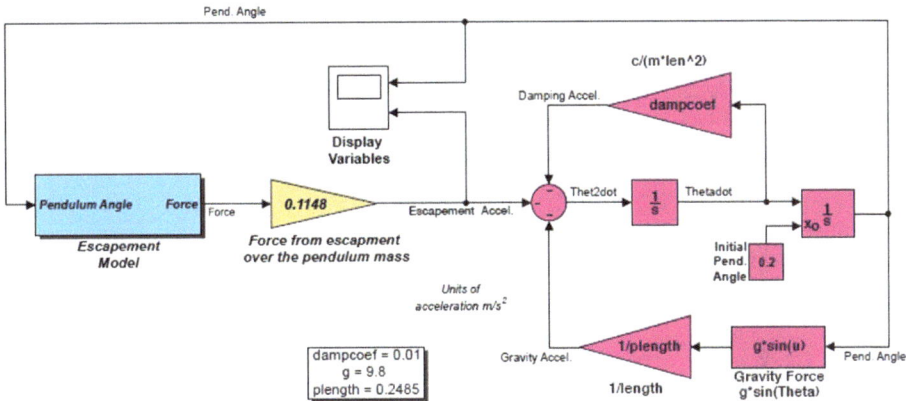

Fig. 1.12 Pendulum clock simulation model

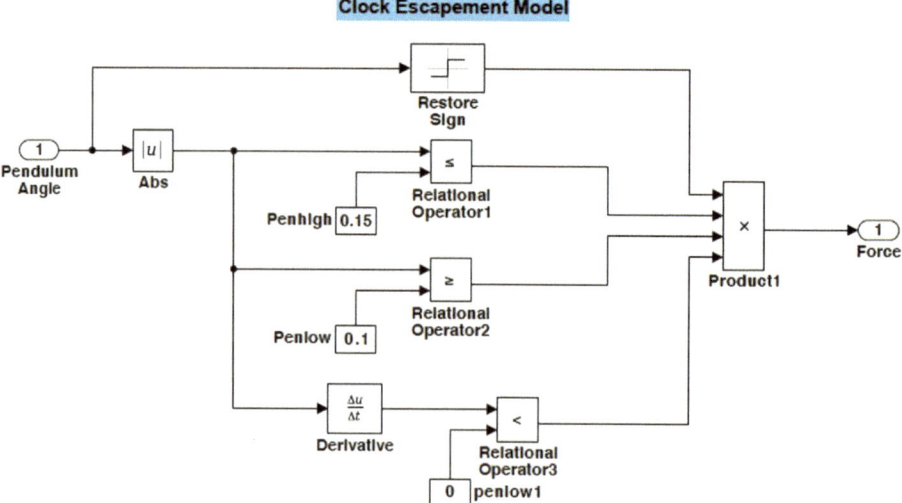

Fig. 1.13 Escapement simulation model

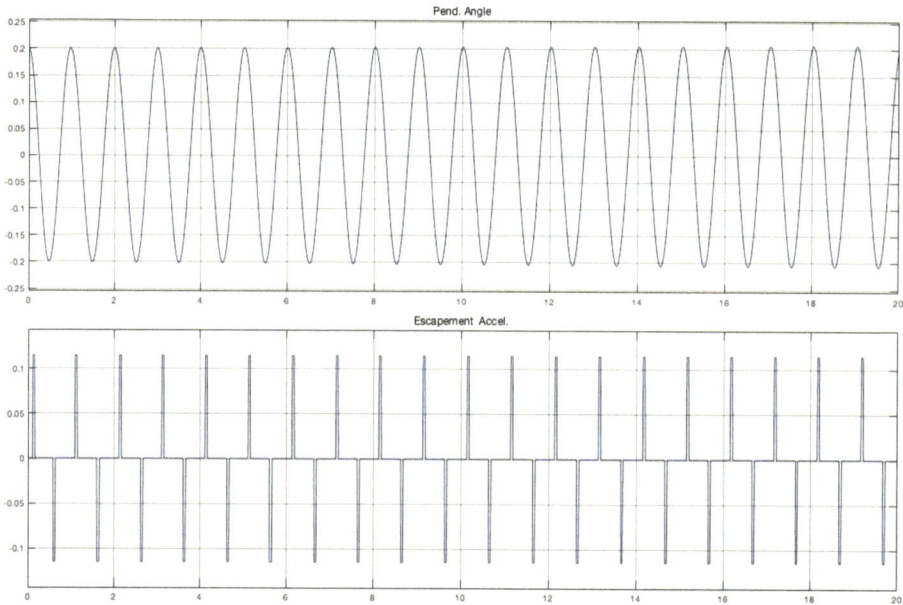

Fig. 1.14 Results of the simulation model

We all have the happy experience of pushing a child on a swing, where, perhaps without realizing it, we function as an escapement. If we want to maintain the amplitude of the oscillation we must do two things: push him at the right time (when the descent begins in the case of the swing), and adjust the force so that it is the strictly necessary to overcome the friction of both the pivot and the body into the air, and to complement the components of the weight that naturally help or hinder the movement depending on its direction. These two actions are represented in our model, respectively, by the escapement operating condition (Eq. 1.2) and by the gain that weights the output.

The mechanical clock, which reached its conceptual plenitude in the century of the Scientific Revolution, can be seen as the first modern machine. First, it combined in its design the causality principle of the science then emerging as well as the mechanist philosophy, and inherited the problem-solving and empirical knowledge characteristic of the engineering. Second, it is, at the same time, the most perfect automaton and the most fruitful metaphor of the functioning of the universe until the advent of the digital era. Third, it fuses energy and information, a fact that defines the modern machines. Finally, because the regular measurement of time is the first manifestation of the «application of quantitative methods of thinking to the study of Nature» and helps to believe in «an independent world of mathematically measurable sequences» [18], the world of the Newtonian science. And so, the clock goes from being an instrument to measure time, natural, telluric or cosmic, and becomes the monarch of time: it produces, governs and organizes a common time to which we must adjust to live in modern society.

References

1. Plato, *República* (Editorial Universitaria de Buenos Aires, 1984)
2. Plato, *Timaeus and Critias* (Penguin, 1987)
3. J.T. Fraser, Out of plato's cave: the natural history of time. Kenyon Rev. **2**(1), 143–162 (1980)
4. E. Panofsky, *Estudios de iconografía* (Alianza, 1985)
5. Heródoto, *Historia, Libros I–II* (Gredos, 1983)
6. A. Gelio, *Noches Áticas* (Porrúa, 1999)
7. E. Jünger, *El libro del reloj de arena* (Tusquets, 1998)
8. O. Mayr, *Authority, Liberty & Automatic Machinery in Early Modern Europe* (The Johns Hopkins University Press, 1989)
9. A.P. Usher, *A History of Mechanical Inventions* (Harvard University Press, 1966)
10. J. Aracil. *Fundamentos, método e historia de la ingeniería* (Síntesis, 2010)
11. D.S. Landes, *Revolution in Time: Clocks and the Making of the Modern World* (Belknap Press, 1983)
12. J. Needham, L. Wang, D.J. De Solla Price, *Heavenly Clockwork: The Great Astronomical Clocks of Medieval China* (Cambridge University Press, 2008)
13. C.A. Ronan, *The Shorter Science and Civilisation in China: An Abridgement of Joseph Needham's Original Text*, vol. 1 (Cambridge University Press, 1997)
14. D. Rooney, *About Time: A History of Civilization in Twelve Clocks* (Viking, 2021)
15. M. Shallis, Time and cosmology, in *The Nature of Time*. Raymond Flood and Michael Lockwood (Basil Blackwell, 1986), pp. 63–79

16. H.A. LLoyd, Timekeepers—an historical sketch, in *The Voices of Time* ed. by J.T. Fraser (George Braziller, 1966), pp. 388–400
17. C. Schwartz, R. Gran, Describing function analysis using MATLAB and Simulink. IEEE Control Syst. Mag. 19–26 (2001)
18. L. Mumford, *Técnica y civilización* (Alianza, 1979)

*In other words, what remains of the movement from which time is
"eliminated"? Is there anything left?*

Alexandre Koyré *Galileo Studies.*

2.1 Galileo, Huygens and the Tautochrone

2.1.1 Galileo

Galileo thought that the period of the pendulum oscillation was independent of the amplitude, that it was only dependent of the length, the radius of the arc described by the swing. When he found a different experimental result, he attributed the difference to the drag force due to the friction of the bob in the air. It was true, but not all the true: the main reason of the difference was that the swing describes a circle, and circular swings are not *isochronous*. So, the clock has a *circular error*. Or, in other words, the period of the pendulum was not independent of the amplitude.

It seems that Galileo built his pendulum clock using an escapement of his own invention instead of the traditional verge, and suspended the pendulum fixing it to a pivoting arbor, and with this kind of escapement he used to experiment with small arcs.

In modern notation, the pendulum equation, discarding friction forces and external stimuli, is:

$$ml\ddot{\theta} + mg\sin\theta = 0, \tag{2.1}$$

where θ is the angle, m the mass of the bob, l the length, and g is the Earth's gravitational acceleration. For small initial angle, θ_0, the sine function can be approximated by

$$\sin\theta = \theta - \frac{\theta^3}{3} + \frac{\theta^5}{5} + \cdots, \tag{2.2}$$

© The Author(s), under exclusive license to Springer Nature Switzerland AG 2024
B. M. Vinagre, *Time in Control Theory*, Synthesis Lectures on Electrical Engineering,
https://doi.org/10.1007/978-3-031-54042-4_2

by the arc if we linearize: $\sin(\theta) \simeq \theta$. In such conditions, the period of oscillation is independent of the initial angle, that is, independent of the amplitude of oscillation, and has the expression

$$T_0 = 2\pi \sqrt{\frac{l}{g}}. \tag{2.3}$$

Therefore, it is easy to understand that Galileo could attribute the differences observed in his experiments with respect to that ideal value to friction. Huygens, defending the originality of his discoveries, argued that «because of that, motion becomes much more difficult and the clock is liable to stop» [1, p. 117], and using a verge escapement, a mechanism that requires an oscillation of wide amplitude, quickly became aware of the circular error and the corresponding lack of isochrony [1].

In this case, the approximation of the sine by the arc is no longer applicable, and it is clearer in experiments that the period of the pendulum depends on the amplitude of the oscillation. And therefore, in practice, when non-zero friction, both in mechanisms and the bob swinging in the air, has its effect, causing its amplitude to decrease over time, the circular error in the measurement of time became more evident. In Fig. 2.1 it is shown the comparison of periods corresponding to 5° swinging and 80° swinging, for both the case of linear approximation and the nonlinear model. It is evident that for 80° the period is larger.

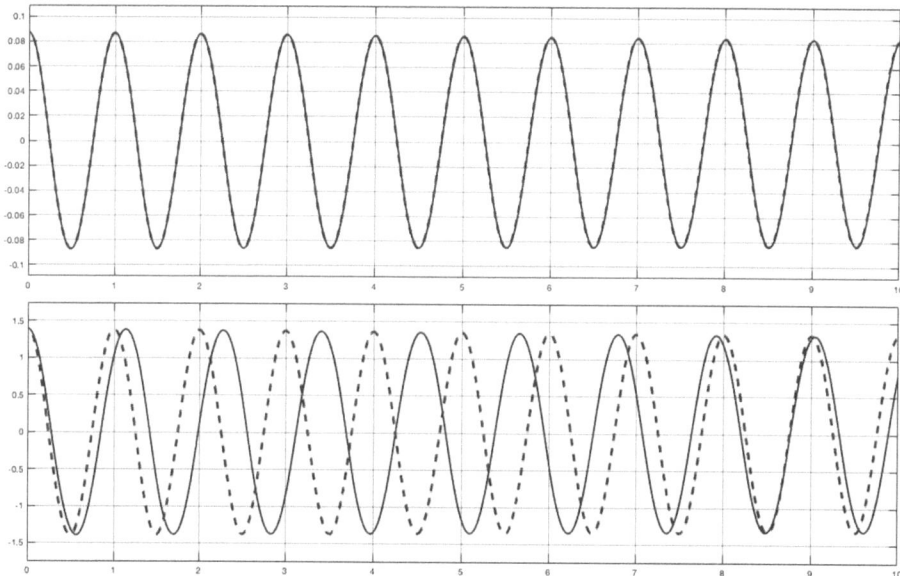

Fig. 2.1 Evolution of the pendulum depending of initial conditions (amplitude of swinging) illustrating the circular error. Up: linearized system. Bottom: nonlinear system. Continuous line: 80° initial condition. Dashed line: 5° initial condition

For obtaining the proper formula for calculating the period of the pendulum, even in the frictionless case, it is necessary to integrate the nonlinear differential Eq. 2.1. Although apparently simple due to the familiarity with the sine function, its solution includes an elliptic integral, and has the form [2]

$$T = \frac{2}{\pi} T_0 K(k),$$ (2.4)

were $K(k)$ is the complete elliptical integral of the first kind defined as

$$K(k) = \int_0^1 \frac{dz}{\sqrt{(1 - z^2)(1 - k^2 z^2)}},$$ (2.5)

being

$$z = \frac{\sin \frac{\theta}{2}}{\sin \frac{\theta_0}{2}}; \quad k = \sin \frac{\theta_0}{2}.$$ (2.6)

This solution can be expressed in a series form [3]

$$T = T_0 \sum_{n=0}^{\infty} \left(\left(\frac{(2n)!}{(2^n n!)^2} \right)^2 \sin^{2n} \frac{\theta_0}{2} \right).$$ (2.7)

being θ_0 the initial angle or, in other words, the amplitude of the swing. So, the period of the pendulum actually increase with the amplitude of swinging, as shown in Fig. 2.1. By using the approximated expression [4]

$$T = T_0 \sqrt{\frac{1}{\cos \frac{\theta_0}{2}}}$$ (2.8)

the Fig. 2.2 represents the evolution of the period with the amplitude of the swing.

2.1.2   Huygens

How to correct this circular error? Huygens devised several possible solutions, among which the one that was successful, although not applicable in practice, was to design a way to tune the effective length of the pendulum during swinging. Once the mechanism was built, a few tests were enough to see that the curve that worked best looked like a cycloid, a curve that is generated by a point on the circumference of a circle that rolls along a straight line without slipping nor friction (Fig. 2.3). Its parametric equations are:

$$x = r(\theta - \sin \theta),$$ (2.9)
$$y = r(1 - \cos \theta).$$ (2.10)

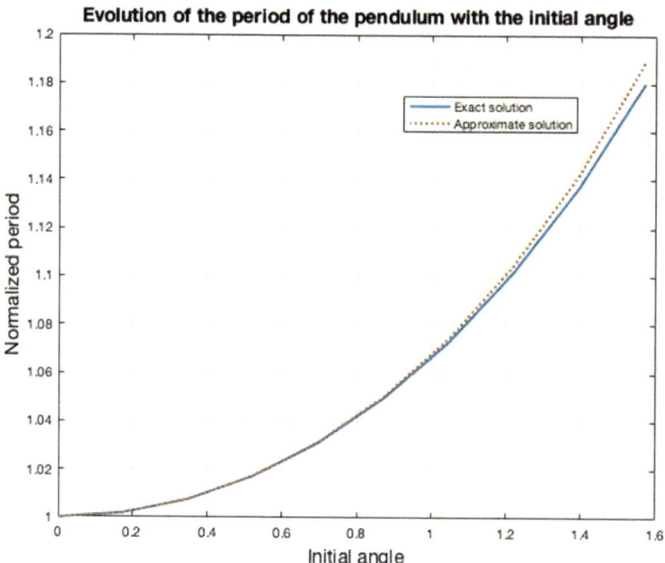

Fig. 2.2 Evolution of the pendulum period depending of initial conditions (amplitude of swinging)

Fig. 2.3 The cycloid curve

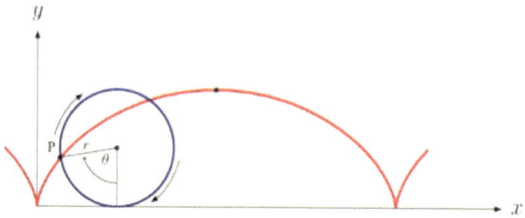

Huygens was not successful in applying a cycloidal swing to the pendulum, since the mechanism used to achieve it imposed limitations on its use (it had to be fixed and preferably vertical), and the solution of putting an intermediate wheel between the verge and the pendulum to avoid requiring a such a wide roll angle added friction and, consequently, irregularity to the gear train. Finally, the solution was a new type of escapement, the *anchor escapement* (Fig. 2.4), named after the appearance of two arms that are responsible, alternatively, for locking and releasing the escapement wheel, and which required an amplitude of much smaller oscillation for proper operation.

But he made an immense contribution to the construction of more precise clocks and, what is more important for our topic, in his *Horologium* he demonstrated geometrically that the proper arc to obtain a isochronous motion was, effectively, the cycloid. He called this curve *tautochrone*, from Greek prefixes *tauto-* meaning same or equal, and *chronos* time.[1]

---

[1] In his total novel *Moby-Dick* Herman Melville also was aware of this fact:«It was in the left hand try-pot of the Pequod, with the soapstone diligently circling round me, that I was first indirectly

Fig. 2.4 Drawing of a recoil or anchor escapement for a pendulum clock, from an 1896 clockmaking manual. The top piece is called the anchor, and the escape wheel is under it

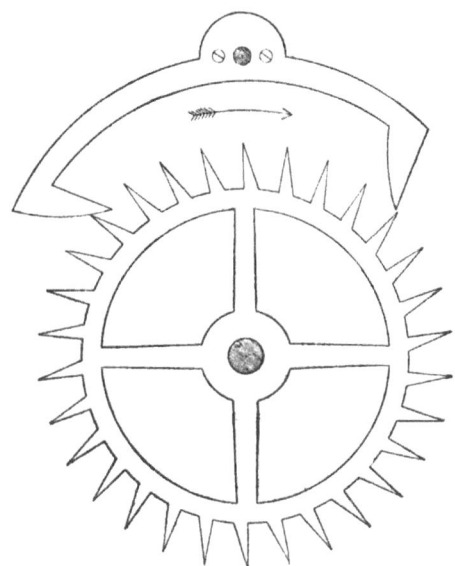

Huygens also proved that the time of descent is equal to the time a body takes to fall vertically the same distance as diameter of the circle that generates the cycloid, multiplied by $\pi/2$. In modern terms, this means that the time of descent is $\pi\sqrt{r/g}$, where r is the radius of the circle which generates the cycloid, and g is the Earth's gravitational acceleration.

2.2   Bernoulli and the Brachystochrone

The geometrical discoveries of Huygens were later used by Johann Bernoulli for solving an, in principle, different problem. In the June 1696 issue of *Acta Eruditorum*, Johann Bernoulli, at this time professor of mathematics at the University of Groningen, The Netherlands, posed the following challenge [5]:

> If in a vertical plane two points A and B are given, then it is required to specify the orbit AMB of a movable point M, along which it, starting from A, and under the influence of its own weight, arrives at B in the shortest possible time.

This challenge is illustrated in Fig. 2.5. Some sixty years earlier, Galileo conjectured that this orbit was a circular arc (Fig. 2.6), and in his challenge Bernoulli stated that «In order to avoid a hasty conclusion, it should be remarked that the straight line is certainly the line of

---

struck by the remarkable fact, that in geometry all bodies gliding along the cycloid, my soapstone for example, will descend from any point in precisely the same time.» (Herman Melville, *Moby-Dick*, Chap. 96, "The Try–Works").

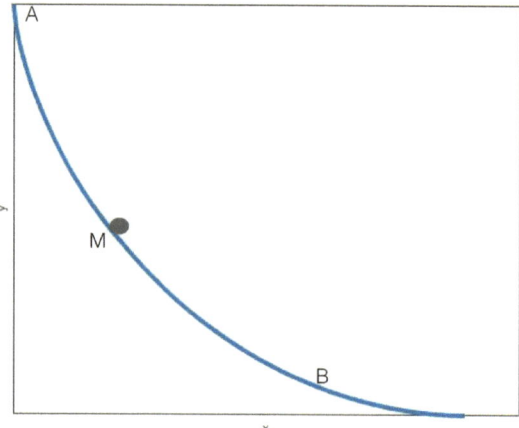

Fig. 2.5 Brachistochrone problem

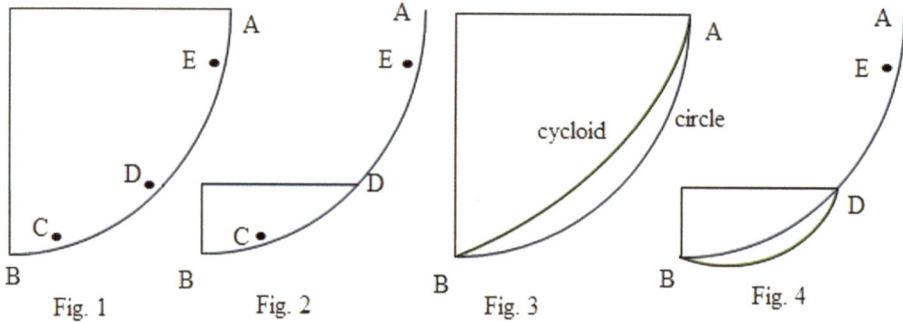

Fig. 2.6 Galileo's shortest time curve conjecture

shortest distance between A and B, but it is not the one which is traveled in the shortest time»
[5]. The solution was, as in the case of the tautochrone, the *cicloyd*, a curve that Galileo
studied and named, but without realizing the connection between it and his problem.

Bernoulli called this orbit of «shortest possible time» *brachystochrone*, from the Greek
$\beta\rho\acute{\alpha}\chi\iota\sigma\tau\sigma\varsigma\ \chi\rho\acute{\sigma}\nu\sigma\varsigma$, meaning 'shortest time', and for being a «true minimum time problem
of the kind that is studied today in optimal control theory», probably determines the beginning
of optimal control theory, so confirming that «it is not, as it may seem, purely speculative
and without practical use. Rather it even appears, and this may be hard to believe, that it is
very useful also for other branches of science than mechanics», as Bernoulli himself stated
[5].

It is worth noting that if the movable point travels at constant speed, that is, if no forces
are involved in the system, the shortest possible time is obtained by following the shortest
distance between A and B, the straight line connecting the two points, and this line is the

particular case of the geodesic for a flat surface in the Euclidean space, $\mathbb{R}^3$. But if the speed depends on the chosen path, then the minimum distance and minimum time problems must be separated.

Curve minimization problems were known at least since the Greeks. The oldest one is determining the shortest path between two points, being the solution a straight-line segment. Other interesting problem was the known as *isoperimetric problem*: to find the plane curve of a given length that encloses the largest possible area. Its solution is, of course, the circumference. Also the Greeks discovered that a light ray that travels from a point A to a mirror and then, reflected, to a point B, could in theory take many paths, but it takes the shortest one. And, if the medium is the same, the shortest path is also the one requiring the least or *minimum* time. If the media traveled are different, then the speeds of light in each medium are in the same relationship as the sine of the angles that the rays form with the vertical, as shown by Descartes and Snell. Fermat later demonstrated that this trajectory is also the one that requires the least time. His *Principle of Least Time* says that «a ray of light travelling from one point to another will always take the path requiring the least time» [6]. This principle played an important role in the solution that Bernouilli gave to the brachystochrone problem [5].

In the challenge posed by Bermoulli, the only acting force is the gravity, but imagine that the moving body is a point in a ray of light. As Bernoulli learned from Galileo, a body falling from height y has a velocity that responds to the expression: $v = \sqrt{2gy}$. Suppose that we change the physical units so that $2g = 1$. Then the problem is exactly equivalent to that of determine the minimum-time paths in a plane medium where the speed of light c varies continuously as a function of position according to the formula $c = \sqrt{y}$. Now, if we divide the plane into very small horizontal strips, we can assume c as a constant in each of them, and therefore we can study this local problem using the law of light refraction. Thus, in each strip the trajectories will be straight lines and what must be determined are the angles they form when crossing the borders of the different strips, applying the laws of Snell, Fermat and Huygens. With this approach, Bernoulli concluded that the relationship between the sine of the angle of incidence and the square root of the height in each strip was constant, which, passing the limit when the number of strips into which we divide the plane tends to infinity, and therefore, setting their corresponding heights to zero means that the angle between the tangent to the curve and the vertical axis must be proportional to $\sqrt{y}$. Therefore, since

$$\sin \theta = \frac{dx}{\sqrt{dx^2 + dy^2}}, \tag{2.11}$$

we find that

$$\frac{dx^2}{dx^2 + dy^2} = ky, \tag{2.12}$$

being k a constant. This relation can be reformulated as the differential equation

$$\left[1 + \left(\frac{dy}{dx}\right)^2\right] y = C,$$

(2.13)

or

$$\dot{y}(x) = \sqrt{\frac{C - y(x)}{y(x)}}.$$

(2.14)

The curves given by parametric equations

$$x(\varphi) = r(\varphi - \sin \varphi),$$

(2.15)

$$y(\varphi) = r(1 - \cos \varphi), \quad 0 \le \varphi \le 2\pi,$$

(2.16)

satisfy (2.14). As we know, these equations corresponds to the cycloid (see Fig. 2.3) generated by a point P that is at the origin of coordinates when $\varphi = 0$, on a circle of radius $r = C/2$ that rolls without slipping nor friction on the horizontal axis. Furthermore, for points A and B of our original problem, there is exactly one curve passing through A and B and belonging to the family of curves (2.16). For given φ, the circle's centre lies at $(x, y) = (r\varphi, r)$. In the brachistochrone problem, the motion of the body is given by the time evolution of the parameter $\varphi(t) : \omega t, \omega = \sqrt{g/r}$, where t is the time since the release of the body from the point $(0, 0)$.

If friction is included, a rather slightly more tangled procedure can be used to obtain

$$x(\varphi) = r[(\varphi - \sin \varphi) + \mu(1 - \cos \varphi)],$$

(2.17)

$$y(\varphi) = r[(1 - \cos \varphi) + \mu(\varphi + \sin \varphi)].$$

(2.18)

In Fig. 2.7 a representation of the time evolution of the arc, s, is given for the case of friction. For the case of no friction, Fig. 2.8 represents the evolution of coordinates x, y with time. In other words, this figure informs of the time to travel from the origin to the point whose coordinates are in the intersection with the corresponding vertical line.

Studying these and other natural phenomena, eighteenth-century scientists realized that nature behaves in such a way that the value reached by some magnitude is either a maximum or a minimum. And to determine these values, which maximize or minimize a related variable, the ideal tool was a recently invented one: the *calculus*.

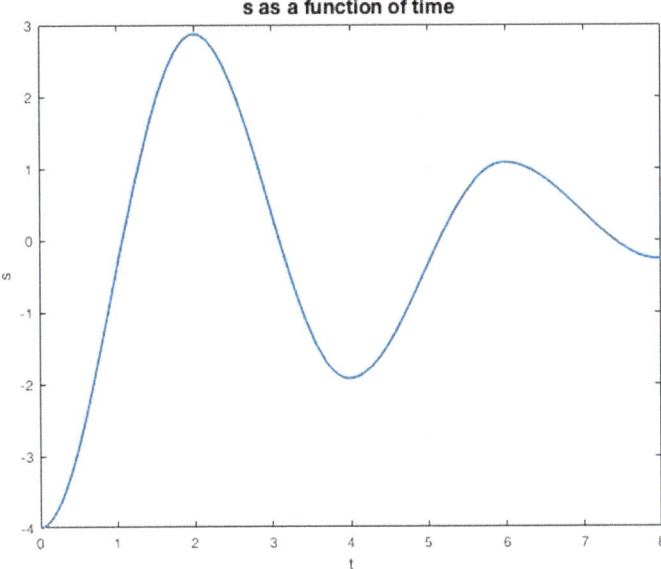

Fig. 2.7 Time evolution of the arc, s

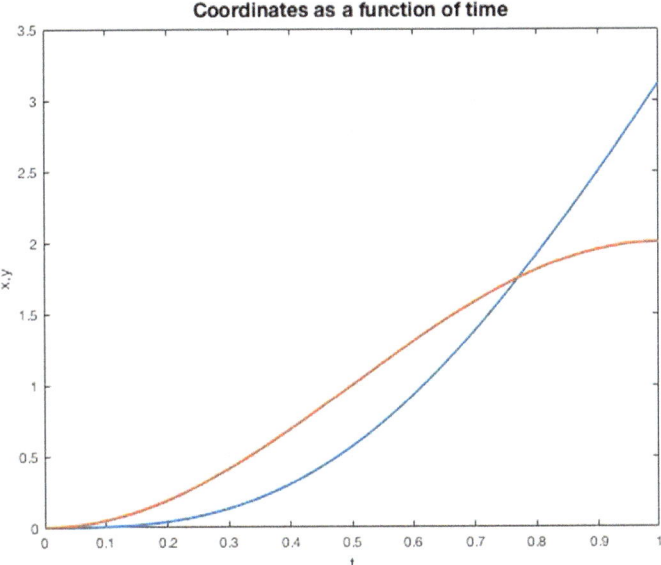

Fig. 2.8 Minimum time

2.3   The Brachystochrone and Other Minimum Time Optimal Control Problems

It must by said that also Galileo formulated the brachystochrone problem, but again as in the pendulum, he thought that the solution was the circle. In both cases, the solution was in fact a curve that himself introduced as *related to the circle*: the cycloid [5]. And this curve is the solution for both, the minimum-time problem and the equal-time problem. Thus, questions about time, when it was minimum or when it was equal, were at the origins of optimization theory, and therefore at the origins of optimal control: this was the beginning of the way followed by Euler, Lagrange or Hamilton, and that after passing the *Calculus of Variations*, has one of its culminating points in the *Pontryagin's Maximum Principle*, to find the best possible control for driving a dynamical system from one state to another, especially in the presence of constraints for the state or control inputs.

Following [5], Johann Bernoulli's original question can be reformulated as an optimal control problem, in which the motion takes place in the x, y plane, and the dynamical behavior of the system is given by

$$\dot{x} = u\sqrt{|y|}, \quad \dot{y} = v\sqrt{|y|}, \tag{2.19}$$

where u, v are the components of a control vector taking values in the set

$$U = \{(u, v) : u^2 + v^2 = 1\}. \tag{2.20}$$

By using Hamiltonian formulation and Pontryagin's maximum principle, we can obtain

$$1 + \dot{y}^2 + 2y\ddot{y} = 0, \tag{2.21}$$

which is a stronger, more elegant and more complete solution to the proposed problem.

Nowadays, optimal control is a branch of modern control theory that deals with designing controls for dynamical systems by minimizing a performance index that depends on the system variables (states, control effort, etc.). There are two basic approaches to optimal control: the calculus of variations, which leads to a formulation of the optimal control solution in terms of partial derivatives of the Hamiltonian; and the *Bellman's Optimality Principle*, which leads to *Dynamic Programming* (DP) [7].

Among the many problems considered in the theory, we are here interested in those concerned with achieving the performance objectives in minimum time. An appropriate performance index for these problems is

$$J = \int_{t_0}^{T} 1\,dt = T - t_0. \tag{2.22}$$

Linear and some nonlinear problems can be explicitly solved by using calculus of variations and hamiltonian formulation, and one of these problems is that of the brachistochrone. In some cases, when for the system

$$\dot{x} = Ax + Bu, \tag{2.23}$$

with $x \in \mathcal{R}^n$, and $u \in \mathcal{R}^m$ is constrained such that $\| u(t) \| \leq 1$ for all $t \in [t_0, T]$, there are difficulties with using the performance index in Eq. (2.22). But adopting a different control strategy, the bang-bang control, the problem can still be solved by applying the Pontryagin's Principle [7].

In the case of discrete-time systems, there is another very interesting problem related to minimum time: forcing the state of a linear discrete-time system to zero in a minimum number of steps. The problem can be formulated as an optimal control problem with no cost on the control. Such controls are called *deadbeat* since they beat the state to a dead stop in at most the same number of periods as the dimension of the state vector [8].

References

1. D.S. Landes. *Revolution in Time: Clocks and the Making of the Modern World* (Belknap Press, 1983)
2. A. Beléndez, C. Pascual, D.I. Méndez, T. Beléndez, C. Neipp, Exact solution for the nonlinear pendulum. Revista Brasileira de Ensino de Física **9**(4), 645–648 (2007)
3. I. Beszeda, T. Stonawski, Pendulum motion on modified trajectories: the cycloidal pendulum as an old-time GPS. Lat. Am. J. Phys. Educ. **15**(4), 4302–1–4302–7 (2021)
4. J.P.J. Neto, Solving the nonlinear pendulum equation with nonhomogeneus initial conditions. Int. J. Appl. Math. **30**(3), 250–266 (2017)
5. H.J. Sussmann, J.C. Willems, 300 years of optimal control: from the brachystochrone to the maximum principle. IEEE Control Syst. 32–44 (1997)
6. M. Kline, *Mathematics in Western Culture* (Pengun Books, 1982)
7. F.K.L. Lewis, Optimal control, in *The Control Handbook: Control System Advanced Methods*, ed. by W.S. Levine (CRC Press, 2011), pp. 25–1–25–35
8. A.E. Naeini, G.F. Franklin, Deadbeat control and tracking of discrete-time systems. IEEE Trans. Autom. Control **27**(1), 176–181 (1982)

Calculus and the Birth of Control Theory

3

In artificial time – ... – the product of segmenting a continuous flow, always different, into a discrete succession of homogeneous instants, the time of the clock – ... –, Achilles does not overtake the tortoise....

Eustaquio Barjau, Preface to Peter Handke's «Poem of the duration».

3.1 The Need for a Tool

3.1.1 From the Workshops to the Study Table

Automatic gadgets were already known in Greco-Latin antiquity, some of the first examples being water clocks or hourglasses described by Vitruvius and attributed to Ctesibius (Alexandria 3rd century BC). In the 1st century of our era, partly based on the knowledge of Ctesibius, Hero of Alexandria describes a great variety of *automata* and mechanisms based on what today we call *feedback* [1]. Since the fourteenth century, and fundamentally through Islamic culture, there has been an interest in the West for the culture of antiquity that gives character to the historical period that today we call the Renaissance. Of course, in the package came, along with philosophy and literature, science and engineering. Throughout the 17th and 18th centuries those old principles gave origin to inventions and new applications especially for temperature control and, as we have seem, timekeeping. The development par excellence of this period was, without a doubt, James Watt's centrifugal governor for his steam engine, whose definitive invention is usually dated to 1789 [2]. But despite the practical success of these devices, Control Theory was yet to be born. How to analyse and design these devices to ensure *stable* operation? How to preserve and transmit the knowledge acquired so that, in a systematic way, the designs could be repeated and improved? All that had been prehistory, and for history to begin, the appropriate language had to be found. This was no other than the mathematical language, which allowed us to jump from the workshops to the study table.

B. M. Vinagre, *Time in Control Theory*, Synthesis Lectures on Electrical Engineering, https://doi.org/10.1007/978-3-031-54042-4_3

3.1.2 The Advent of *Calculus*

At the end of the 16th century, European mathematicians, familiar with Greek mathematical developments and their geometric methods and also with Islamic algebra, merged these two traditions to take two fundamental steps: the invention of *symbolic algebra* by François Viète in 1591, and of *analytic geometry* by Descartes and Fermat around 1630. The first allowed solving geometric problems and the second established that curves could be represented by equations and that each equation determined a curve. Thus, although most of the interests came from the Greeks (tangents, areas, maxima and minima), the object of study (curves and geometric figures) had become widespread, as we have seen studying the cycloid, and the solution lay in these two new tools.

In Chap. 2 we have discussed an example of finding a minimum, to which Fermat's method to find maxima and minima could be applied. But, for us now, even more interesting than the problem of finding extrema, is the one of finding tangents, and to realize that, by the middle of the seventeen century, without the aid of calculus, Fermat, Descartes, Wallis or Barrow, given the equation of a curve, $y = f(x)$, were able to find the tangent by conceiving and computing the slope of the secant using the expression (see Fig. 3.1)

$$\frac{f(x+h) - f(x)}{h},$$

as h becomes smaller and smaller, that is, as the points defining the secant came closer and closer. And Galileo was able to study uniformly accelerated motion supported by the Aristotelian concept of *change*. But they were Leibniz and Newton who, at the end of the 17th century, gathered all the previous baggage under the fundamental concepts of *derivative* and *integral*, who produced a notation that made easy to work with such concepts, and those who gave arguments to prove the *Fundamental Theorem of Calculus*: integral and derivative are inverses. In short, it was they who invented the calculus, although Euler and Lagrange

Fig. 3.1 Computing the tangent

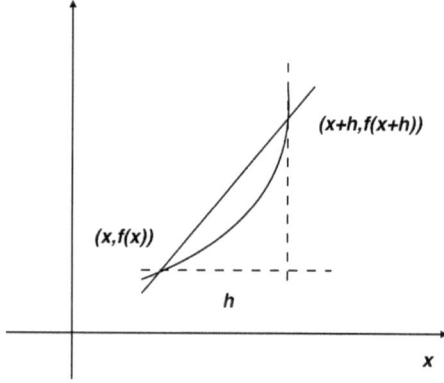

P. Fermat
1637-1638

R. Descartes
1637-1638

I. Newton
1669

G. W. Leibniz
1684

Leonard Euler
1755

J. L. Lagrange
1797

A. L. Cauchy
1823

K. Weierstrass
1861

Fig. 3.2 Great milestones in the development of calculus

still had to come to establish the theory of differential equations, and Cauchy and Weierstrass to provide the definitions and proofs that are today the bases of this mathematical discipline (see Fig. 3.2) [3].

But what we are now interested in underlining is not the later development of calculus, but the philosophical background that underlie the parallel inventions of Leibniz and Newton. This background is already present in the different language that each of them uses: while for Leibniz the derivative is a ratio between infinitesimal differences, a *differential quotient*, for Newton it is a *fluxion*, a rate of flow or change [4–6].

3.2 The Calculus and the Invention of Mathematical Time

3.2.1 Leibniz

In 1684 and 1686, in the aforementioned journal *Acta Eruditorum*, Leibniz published the works in which what we would know as *Mathematical Analysis* appears for the first time. The abbreviated titles of these works were *Nova methodus* and *De Geometria recondita*[1] (see Fig. 3.3).

In the first work, Leibniz defines geometrically the concept of *difference*, dx, and establishes the rules for operating with them: difference of a constant quantity, product by a constant, addition and subtraction of differences, product, division, potentiation and roots. Given this set of rules he finally states that «From the knowledge of this *Algorithm*, so I call it, or from this calculus, which I call differential, all the other differential equations can be obtained by means of common algebra, and the maxima and minima, just as the tangents can be obtained, in such a way that it is not necessary to separate the fractions or irrational or other links, as, however, had to be done according to the methods published until now.» [4]. Among the applications of his method that he includes in the paper is the obtaining of the

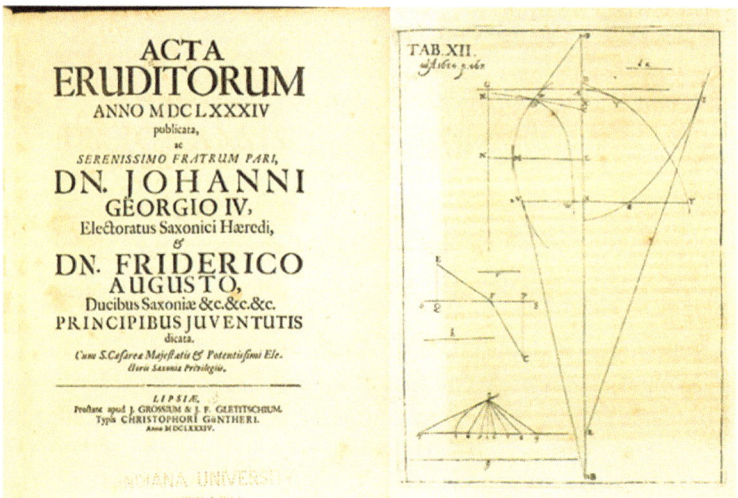

Fig. 3.3 The title page for the 1684 volume of *Acta Eruditorum*, and a plate of diagrams of Leibniz's article *Nova methodus* (https://archive.org/details/s1id13206500)

---

[1] The complete titles are: *Nova methodus pro maximis et minimis, itemque tqngentibus, quae nec fractas nec irrationales quantitates moratur, et singulare pro illis calculi genus* and *De geometria recondita et Analysi indivisibilium atque infinitorum*, something like *A new method for maximums and minimums, as well as for tangents, which does not stop at fractional or irrational quantities, and is a unique type of calculus for these problems* and *On a highly occult geometry and the analysis of indivisibles and infinities*.

law of refraction, the one used by Bernoulli in his brachystochrone problem. He concludes this example by saying that «The expert of this Calculus has obtained in three lines what other very learned men have investigated with many difficulties» [4]. He called the derivative *differential quotient* and used the notation dy/dx. In the second work, which deals with the quadrature of curves, he introduces, for the first time in history, the sum or quadrature as the inverse of the difference, and with the same symbol that we use today, $\int y dx$.

3.2.2 Newton

Consider the reader the curve, y, under which the area up to a certain point $D(x, y)$ is $z = x^3$. In Fig. 3.4 Bb has length o, and v is chosen so that area $BbHK = ov =$ area $BbdD$. Then, if x increases to $x + o$, the area changes from $z(x)$ to $z(x + o)$. So, the change in the area is

$$z(x + o) - z(x) = (x + o)^3 - x^3 = 3x^2 o + 3xo^2 + o^3. \tag{3.1}$$

On the other hand, by definition of the problem, we know that this change in area is equal to ov. So, $3x^2 o + 3xo^2 + o^3 = ov$, and, dividing by o we obtain $3x^2 + 3xo + o^2 = v$. If we now make the point b to be infinitely close to B, v tends to be equal to y, o vanish to 0, and every term containing it will vanish to 0. The final result can be expressed as

$$z(x + o) - z(x) \rightarrow y = 3x^2, \quad o \rightarrow 0. \tag{3.2}$$

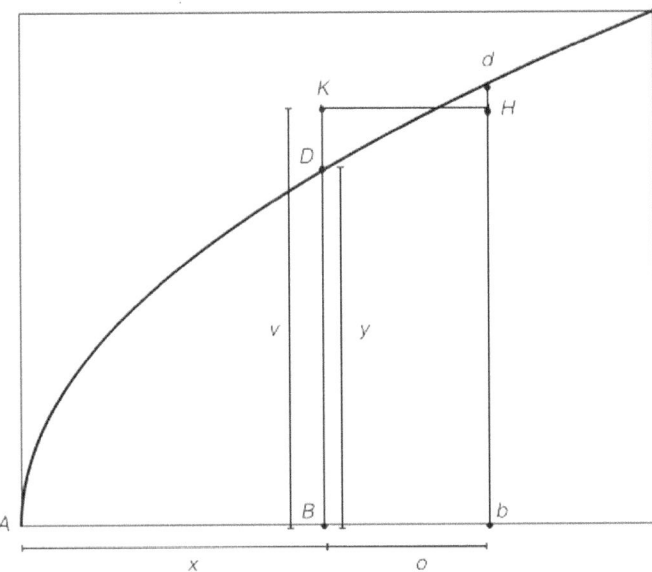

Fig. 3.4 Derivative as *rate of change* (adapted from [3])

In this example, adapted from Newton in [3], we must remark that when $o \to 0$

$$\frac{z(x + o) - z(x)}{o} = y,$$ (3.3)

is the *rate of change* of the area, as well as $3x^2$ is the slope of the tangent to the curve $z = x^3$. And it shows that the new tool can be applied to elegantly solve problems of both, areas and tangents, that these problems are mutually inverse, as the integral and the derivative are.

3.2.3 Leibniz and Newton on Time

Though both, Leibniz and Newton, consider geometrical problems, and though both worked also on dynamics, the fact that for Leibniz the derivative was a ratio between infinitesimal differences, a *differential quotient*, and for Newton a *fluxion*, a rate of flow or change, was crucial. Language matters.

What is exactly a differential quotient? Because things are not necessarily equivalents to the ways they are calculated. For instance, this confusion led the Marquis de L'Hôpital, a disciple of Johann Bernoulli, to ask Leibniz «What if n be $1/2$?» in the expression $d^n y/dx^n$, to which the latter responded [7]:

> You can see by that, sir, that one can express by an infinite series a quantity such as $d^{1/2}\bar{x}y$ or $d^{1:2}\bar{x}y$. Although infinite series and geometry are distant relations, infinite series admits only the use of exponents which are positive and negative integers, and does not, as yet, know the use of fractional exponents. [...] Thus it follows that $d^{1/2}x$ will be equal to $x\sqrt{dx : x}$. This is an apparent paradox from which, one day, useful consequences will be drawn.

This confusion caused by language was the seed of what is known today as *Fractional Calculus* (see Appendix).

And what exactly means that the quantity o used in the later example *vanishes*, as Newton stated? He tried to solve, or avoid, this problem by saying that the quantities defined as $3x^2$ was in the example, are *limits* of the ratios of vanishing increments, or *ultimate ratios*: «the value of the ratio of those vanishing quantities just when they are vanishing» [3]. And later he stated that

> Quantities and the ratios of quantities which in any finite time converge continually to equality, and before the end of that time approach nearer to each other than by any given difference, become ultimately equal.

But let us leave the calculus fighting for its definition, and let us focus our attention on that «in any finite time converge continually». This continuous convergence in any finite time, that Newton will establish in the *Method of Fluxions* and will be one of the mathematical bases of his *Principia* requires a time that «in itself and by its nature flows uniformly and without relation to something external». That is, it needs for the *absolute* or *mathematical* time, a time that is a fundamental part of the structure of the universe, a dimension independent of events in which they occur in sequence; that form, together with space, the scenery of physics [5, pp. 643–1019]. For Leibniz, instead, time was part, together with space and number, of the foundations of our intellectual structure, the one with which we sequence and compare events; time is not measurable, it does not have an existence independent of material objects and processes, it exists only by virtue of the existence of matter and material events. This is the *relational* idea of time, inadequate to initiate the development of dynamics that, in these early moments, needed the *absolute* idea of time postulated by Newton. The story is worth telling, even if briefly.

During 1715 and 1716, Caroline, Princess of Wales, acted as an intermediary between Leibniz and Samuel Clarke, a friend of Newton. Leibniz wrote to her lamenting the decay of natural religion in England, suggesting that, at least in part, Newton was one responsible. The Princess send the letters to Clarke who answered as being the Newton's voice [8]. The correspondence almost became a philosophical bestseller in the 18th Century [9]. Several topics are debated along the letters, and among them there is a detailed account of Leibniz's philosophy of time. Clarke, as Newton, considered time, as space, absolute and existing by themselves. But Leibniz, by using arguments based on both, the *principle of identity of indiscernibles*[2] and the *principle of sufficient reason*,[3] considered that the location of an object is not a property of an independent space, but a property of the object itself. That is, objects can be different simply by virtue of its different locations, being that location a property of the object. Leibniz deals with time from the point of view of space, and then also consider time as intrinsic to the events, the *things in time*[4] [8]: a change in time is a change in the event. So, location is not absolute, but a situation of an event relative to other events. Furthermore, time, as space, is not real, but ideal. Of course, they symbolize real differences between things, but they are not real things, and spatial and temporal relations are not irreducibly exterior to things or events: in brief, «time is nothing over and above an ordered system of events and [...] space was nothing over and above a system of bodies» [8]. Thus, it could be said that Leibniz translates temporal expressions with terms such as beginning, end or instant, in such a way that they only refer to events or temporal relations between events: event without previous events, event without subsequent events, or set of concurrent events.

---

[2] The Identity of Indiscernibles is usually formulated as follows: if, for every property F, object x has F if and only if object y has F, then x is identical to y. Or in the notation of symbolic logic: $\forall F(Fx \leftrightarrow Fy) \rightarrow x = y$. (Stanford Encyclopedia of Philosophy).

[3] Among the many formulation we can cite: (1) For every fact F, there must be a sufficient reason why F is the case; (2) For every x, there is a y such that y is the sufficient reason for x (formally: $\forall x \exists y \, Ryx$ [where Ryx denotes the binary relation of providing a sufficient reason]). (Stanford Encyclopedia of Philosophy).

[4] Events or processes are equivalents to objects, and moments or instants are equvalents to locations [8].

Of course, this conception of time did not satisfy the *Spirit* of an age which had managed to build clocks so precise that they were started to impose the conception of a universal time. To continue advancing in that direction, the Newtonian conception of time, as established in his *Principia* of 1687, was needed [5, p. 655]:

> Hitherto I have laid down the definitions of such words as are less known, and explained the sense in which I would have them to be understood in the following discourse. I do not define time, space, place, and motion, as being well known to all. Only I must observe, that the common people conceive those quantities under no other notions but from the relation they bear to sensible objects. And thence arise certain prejudices, for the removing of which it will be convenient to distinguish them into absolute and relative, true and apparent, mathematical and common.

> I. Absolute, true, and mathematical time, of itself, and from its own nature, flows equably without relation to anything external, and by another name is called duration: relative, apparent, and common time is some sensible and external (whether accurate or unequable) measure of duration by means of motion, which is commonly use instead of true time; such as an hour, a month, a year.

3.3   The Birth of Control Theory

With the Newtonian time «hammered» into the minds [10], with the pendulum clock invented and studied, and with calculus on its way to defining and establishing itself as a rigorous mathematical field, the path to the emergence and development of the Control Theory was paved.

As we saw in Chaps. 1 and 2, in the second half of the 17th century, Christiaan Huygens was successful in finding the cycloidal pendulum to obtain a period independent of the amplitude. To make the pendulum bob follow a cycloidal arc, he suspended it from a string attached at its pivot to the junction of two strips of metal curved as cycloids (see Fig. 3.5), so that when the pendulum moves to either side the string wraps itself along one of the strips and the bob describes a cycloid[5] [11]. But also realized that a clock with and escapement is a kind of speed control device, through working in discrete steps. To avoid discrete steps and make the motion continuous, Huygens designed a new kind of *conical pendulum* (Fig. 3.6) such that to fulfil the condition of a period independent of the height of the circle described by the bob in the spherical surface. This device can be considered as a *governor* for speed regulation with feedback via friction in the bearings: the greater the circle described by the bob, the greater the centrifugal force supplied by the bearings, which results in greater friction and requires a greater driving torque. And so, «Huygens speed control system represents probably the first published design of a controller deliberately intended to eliminate static

---

[5] The evolute of the cycloid, the geometric locus of its centres of curvature, is also a cycloid.

Fig. 3.5 Huygens cycloidal
pendulum

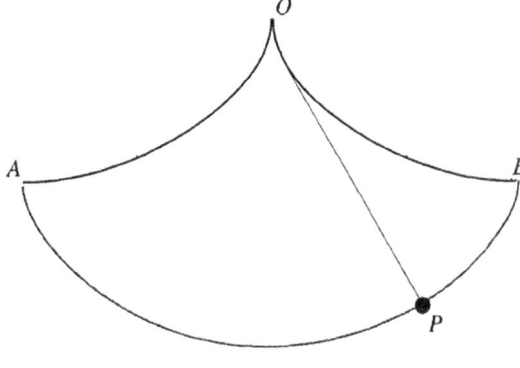

Fig. 3.6 Conical pendulum

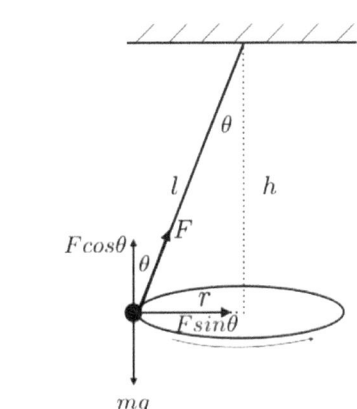

error (offset) due to constant disturbances. [...] Huygen's device is a forerunner of this
technique, though he himself would not have had the notion of integral control» [11]. Of
course, to argue in these terms, notions of mass, inertia, or torque, the Newton's laws, and
the differential equations had to come.

Robert Hooke, the reputed inventor of the anchor escapement, demonstrated the conical
pendulum in 1666, and in 1674 he published a work on the application of it to regulate the
speed of a telescope [11]. He designed a weight-driven equatorial telescope regulated by this
mechanism, in order to track a star smooth and continuously. Three years later, he described
a centrifugal governor with two fly-balls and springs. By doing so, he anticipated both, the
problem of guidance and the centrifugal governor used by James Watt.

But it was the problem of the stability of the movement, brought about above all by the
interest and need to design better *governors*, making use of the knowledge accumulated in
the design of pendulum clocks, which gave birth to control theory. George Biddell Airy,
Astronomer Royal at Greenwich Observatory, published in 1840 a paper on the dynamics
of governors for astronomical telescopes. In it, he studied the dynamics of several kinds
of governors and showed how the motion of them could be described using *differential*

equations based on the pendulum model (1840 and 1851). Unfortunately, he did not succeed in determining the stability conditions. It was not possible until James Clerk Maxwell published his famous article *On Governors* in 1868 [12]. In it he describes how to derive the differential equations for various types of regulators and how to determine their stability from the location of the roots of the *characteristic equation*. Later, Edward John Routh found a criterion to determine the stability of systems of any order (1877), servomechanisms were developed, electrical engineering and operational calculus were incorporated for his study. It came the telegraph, and the telephone, the frequency domain and the feedback theory [2].

But this story is well known, and can be explored in [2, 13–15] and many other sources. What is interesting for our subject is that, having been the Newtonian time the facilitator of these developments, time as a problem disappeared during the successful stage of classical control untill around 1950. Walter Evans had already noted the need for returning to the domain of time with his *Root Locus* (1948) [16, 17], but the contributions of John Ragazzini and his disciples Eliahu Jury, Rudolf Kálmán and Lofti Zadeh where the ones that allowed the start of a new stage that would bring us, among other things, the *digital control* and the *control in the state space*, a stage that would highlight back the time in the foreground.

And just around the corner, Leibniz's conception of time, with its events and durations, was waiting for us.

References

1. A. Aracil, *Juego y artificio: Autómatas y otras ficciones en la cultura del Renacimiento a la Ilustración* (Cátedra, 1998)
2. S. Bennett, *A History of Control Engineering 1800–1930* (Peter Peregrinus—IEEE, 1979)
3. J.V. Grabiner, The changing concept of change: the derivative from Fermat to Weierstrass. Math. Mag. **56**(4), 195–206 (1983)
4. G.W. Leibniz, *Análisis Infinitesimal* (Tecnos, 1994)
5. I. Newton, Principios matemáticos de la filosofía natural, in *A hombros de Gigantes*, ed. by S. Hawking (Crtica, 2003), pp. 643–1019
6. M. Kline, *Mathematics in Western Culture* (Pengun Books, 1982)
7. Bertrand Ross, The development of fractional calculus 1695900. Hist. Math. **4**(1), 75–89 (1977)
8. W.H. Newton-Smith, Space, time and space–time: a philosopher's view, in *The Nature of Time*, ed. by R. Flood, M. Lockwood (Basil Blackwell, 1986), pp. 22–35
9. G.W. Leibniz, S. Clarke, *Correspondance* (Hackett Publishing Company, 2000)
10. T. Damour, Time and relativity, in *Time- Poncaré Seminar 2010*, ed. by B. Duplantier (Birkhäuser, 2013), pp. 1–19
11. A.T. Fuller, The early development of control theory. J. Dyn. Syst., Meas., Control Trans. ASME **98**(2/June), 109–118 (1976)
12. James C. Maxwell, On governors. Proc. R. Soc. **100**, 270–283 (1868)
13. S. Bennett, *A History of Control Engineering 1930–1955* (Peter Peregrinus—IEEE, 1993)
14. R. Bellman, R. Kalaba, *Classic Papers in Control Theory* (Dover Publications, 2017)

15. T. Başar, *Control Theory: Twenty-Five Seminal Papers (1932–1981)* (IEEE Press, 2001)
16. Walter R. Evans, Graphical analysis of control systems. Trans. AIEE **67**(1), 547–551 (1948)
17. Walter R. Evans, Control systems synthesis by root locus method. Trans. AIEE **69**(1), 66–69 (1950)

Sampling and Integral Sums

*In short, the practically cognized present is no knife-edge, but a
saddle-back, with a certain breadth of its own on which we sit
perched, and from which we look in two directions into time.*

William James, The Principles of Psychology

For systems ruled by differential equations, the underlying time was that Newtonian time
that flows continuously and uniformly and can take values along the entire line of real
numbers: between any two instants there is an infinite number of time points, and the
variables considered have a certain value at a certain instant or infinitesimally short duration.
Digital control, the control that can be implemented in a digital computer (see Fig. 4.2),
means abandoning at least one of these properties of time: although in the most simple case
it remains uniform, its flow is not continuous but consists of a sequence of distinct and
separate points at each of which the variables, which have remained unchanged between
two consecutive ones, experience a jump. In this way, once a period T has been set to
take samples of the variables, each instant can be identified with an index that is always an
integer, and the variables are represented by a sequence of values. Without now going into the
property or impropriety of the denomination *discrete time* or the physical and philosophical
considerations of such a conception of time, I want to focus on the *sampling* itself.

Taking into account the relationship between successive sampling instants, we can consider at least three ways of sampling a signal: synchronous, semi or quasi-synchronous and
asynchronous (Fig. 4.1). Synchronous sampling means that samples are taken at equidistant
moments in time, periodically. In many practical cases, however, this procedure is not fully
applicable and only an average period can be guaranteed for the samples, varying the actual
period within certain limits. An alternative to this *time-based* sampling (and control) is to
sample when it is 'worth it'. This brings us to asynchronous *event-based* sampling and
control [1].

B. M. Vinagre, *Time in Control Theory*, Synthesis Lectures on Electrical Engineering,
https://doi.org/10.1007/978-3-031-54042-4_4

Fig. 4.1 Forms of sampling

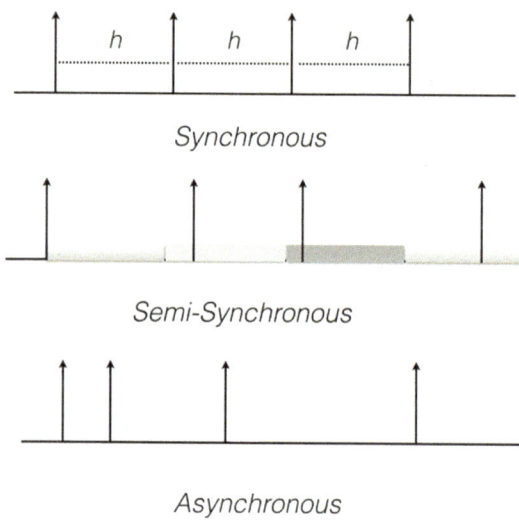

4.1   Time Abandons Continuity

When we want to do digital control of a naturally continuous system we need to do two things: use the theory of *sampled data systems*, continuous systems whose inputs and outputs are sampled, for obtaining a *discrete equivalent* of the continuous system, and use an algorithm or numerical technique to execute the control law. Both of these task assume a basic idea, the idea of *numerical integration* to solve a differential equation.

The most common way of designing digital control systems is to use periodic sampling, that is, to sample the signals at equidistant moments of time. Thus the analysis and design are simpler: for time-invariant linear systems the closed-loop system is linear and periodic,

Fig. 4.2 Basic scheme of a
computer controlled system

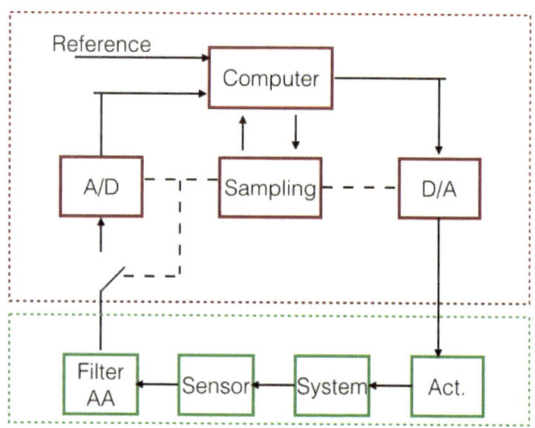

the time-varying nature of the sampled systems disappears and the system can be described by constant coefficient difference equations, and furthermore this fits perfectly with the way computers work and the clock-ruled model of real-time software [2].

The numerical integration technique that supports this approach is that of the Riemann, in which to calculate the definite integral of a function $f(t)$ in the interval $[a, b]$, conceived as the area under the curve between those points, we proceed to partition the interval into subintervals, $[t_k, t_{k+1}]$, such that

$$a = t_0 < t_1 < t_2 < \cdots < t_i < \cdots < t_n = b,$$

obtaining then an approximation to the required integral as a Riemann sum (Fig. 4.3), the sum of areas of rectangles with bases equal to the width of the corresponding subinterval and height equal to the values of the function at a given point which lie in it, $x_k \in [t_k, t_{k+1}]$ for each k. That is:

$$\int_a^b f(t)dt \simeq \sum_{k=0}^{n-1} f(x_k)(t_{k+1} - t_k) = \sum_{k=0}^{n-1} f(x_k)\Delta t. \tag{4.1}$$

The size of the subintervals, Δt_k, would give a result more or less precise in the computation of the integral, with the selection of the intermediate points in each subinterval which, in addition, generate the particular *rule* used for approximating the area of each subinterval, as the forward Euler and the backward Euler rules.

This fits perfectly with that «absolute» Newtonian time that «flows uniformly and without relation to anything external», the time that allowed the extraordinary development of science and technology in the last four centuries. But

Through which organ do we perceive time? ...... In order to measure it, it would be necessary for it to pass in a uniform way. Where is it written that it does it? It doesn't give us that feeling, of course, we only accept that it does it to guarantee order, ... [3].

Fig. 4.3 Riemann integral sum

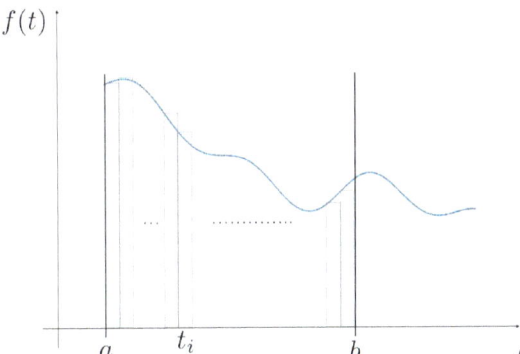

4.2   Time Abandons Uniformity

For systems complex enough, we can have the necessity of using different rates for updating different inputs and outputs. This gives to the systems a time-varying nature, and can characterise complexity in the sense of concurring different time scales. In some cases we can simplify the analysis and design problems if we assume that all sampling operations are synchronised at periodic intervals, and all sample rates are integer multiples of some common or fundamental rate. But we are now interested in the cases in which these conditions are not fulfilled, when having not multirate but totally asynchronous sampling.

Let's accept for the time being that time «pass in a uniform way» and that we have sufficiently precise instruments to measure it. Let's choose the most precise, an atomic clock and define, with the International System of Units, that a second is

> The duration of 9 192 631 770 oscillations of the radiation emitted in the transition between the two hyperfine levels of the ground-state of the isotope 133 of the caesium atom at a temperature of 0 K.

This a very bizarre definition, one that leaves us somewhat perplexed: at the extreme of precision we find that the unit of time is defined by an event, that the time perceived by clocks has no more reality than which is perceived by us. Actually, the pendulum to which we have referred so much also marked time in jumps governed by the escapement. However, these jumps are, or are intended to be, equidistant, marking a fixed rhythm, like the synchronous sampling that we have mentioned. To calculate the integral, we are using Riemann sums.

But we can calculate the area under the curve by taking horizontal strips instead of vertical ones. By doing so, we change from Riemann to Lebesgue for calculating the integral and, concerning sampling, there is no clock that marks the sampling instants, but these and the intervals they define are determined by the values or increments of the function that is being integrating, as expressed in Eq. (4.2) and shown in the Fig. 4.4.

$$\int_a^b f(t)dt \simeq \sum_{k=0}^{n-1} f_k(\Delta t)_k \tag{4.2}$$

This is the approach of the *event-based control*, which has forced new theoretical developments and still has a long way to go, but which allows optimizing the solution of some problems very common in modern control: hybrid control, discontinuous control, distributed control, or networked control. In addition, it is the mode of operation of devices for general use, such as encoders, on-off actuators or intelligent sensors, and the control strategy (the control is not executed until it is required) has conceptual and practical advantages, and it turns out to be an interesting mix of feedback and feedforward control often found in biological systems [1, 2, 4].

Fig. 4.4 Lebesgue integral sum

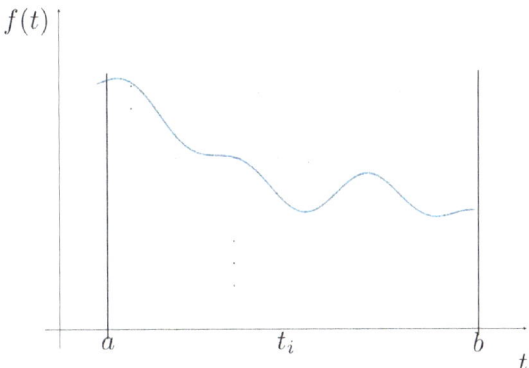

4.3   Leibniz Relational Time Revisited

Past youth we begin to perceive how memory is selective, that it is not a continuous, uniform and non-specific record of everything that happens to us, that we forget most of the things that happen to us from Monday to Friday but we remember that trip, that party or that face among a sea of faces. Our memory, as an integral, does not follow the Riemann method. The *relational* idea of time defended by Leibniz reappears: time exists only by virtue of events.

We get closer to the experience if we consider that «from a purely empirical point of view, the idea of 'event'–and the relationship between events–is more fundamental, and that our normal concepts of space and time are to be understood as idealized mathematical constructs of empirical data that correspond to the events», and that dynamics could be formulated in a new way in which the concepts of «becoming» and «change» are independent of any underlying temporal being [5].

Event-based sampling that accommodates Lebesgue integral sums is called *level crossing sampling*, but other alternatives have been studied, such as *integrated level crossing* sampling or the sampling by *level crossing + linear prediction*. In the first case, the sample is updated if the integral of the absolute value of the difference between the current value of the signal and the last sample reaches a certain threshold. In the second, it does so if the difference between the current value and the one obtained through linear prediction reaches said threshold [1]. In both cases, the event that triggers sampling is not the mere value of the function $f(t)$ but that of a function related to it or to the independent variable that we can generalize as $g(f(t))$.

We can go one step further by resorting to another type of integral, that of Stieltjes. The most common definition is

$$\int_a^b f(t)dg(t) \simeq \sum_{k=0}^{n-1} f(c_k)(\Delta g)_k, \tag{4.3}$$

where $f(t)$ and $g(t)$ are real functions of the real variable t, $(\Delta g)_k = g(t_{k+1}) - g(t_k)$, and c_k is in the subinterval $[t_k, t_{k+1}]$. The function $g(t)$, often called the integrator, can play several roles.

The most obvious is that of a weighting: the values of $f(t)$ do not all have the same weight, either because of their value, content or meaning, or because of the instant in which they are produced, either by both.

$$\int_a^b f(t)g(f(t))dt; \quad \int_a^b f(t)g(t)dt \tag{4.4}$$

Two examples are shown in the Fig. 4.5. One of them wants to represent how events lose importance as they move away from the present, how the memory, the trace they leave in memory, fades as they go into the past; the other, how the meaning we give to events determines the trace they leave in memory.

Now suppose that the function $g(.)$ has the form [6]

$$g(\tau) = (t^\alpha - (t - \tau^\alpha)), \quad 0 \leq \tau \leq t, \tag{4.5}$$

that we see represented in Fig. 4.6 up on the left. On the right we can see the three-dimensional representation of the functions g and f as a function of the independent variable τ. The two figures on the bottom part represent the projections of those curves on the (τ, f) (left) and (g, f) (right) planes, and the áreas corresponding, respectively, to the integrals

$$\int_a^b f(t)dt, \quad \int_a^b f(t)dg(t). \tag{4.6}$$

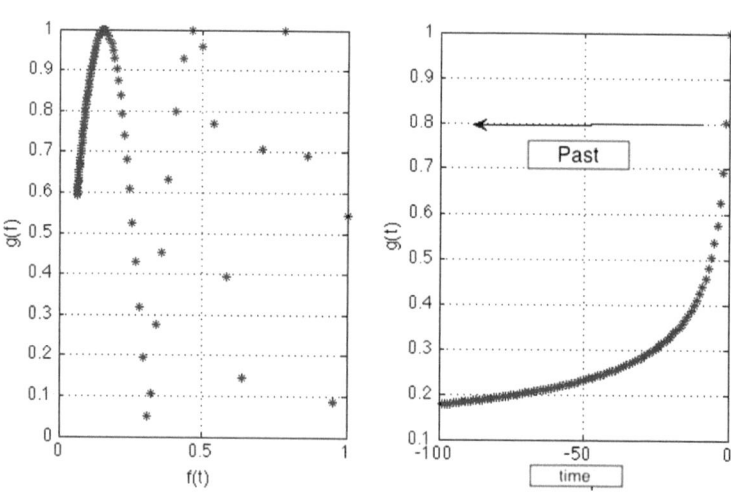

Fig. 4.5 Something like memory: weights of the events

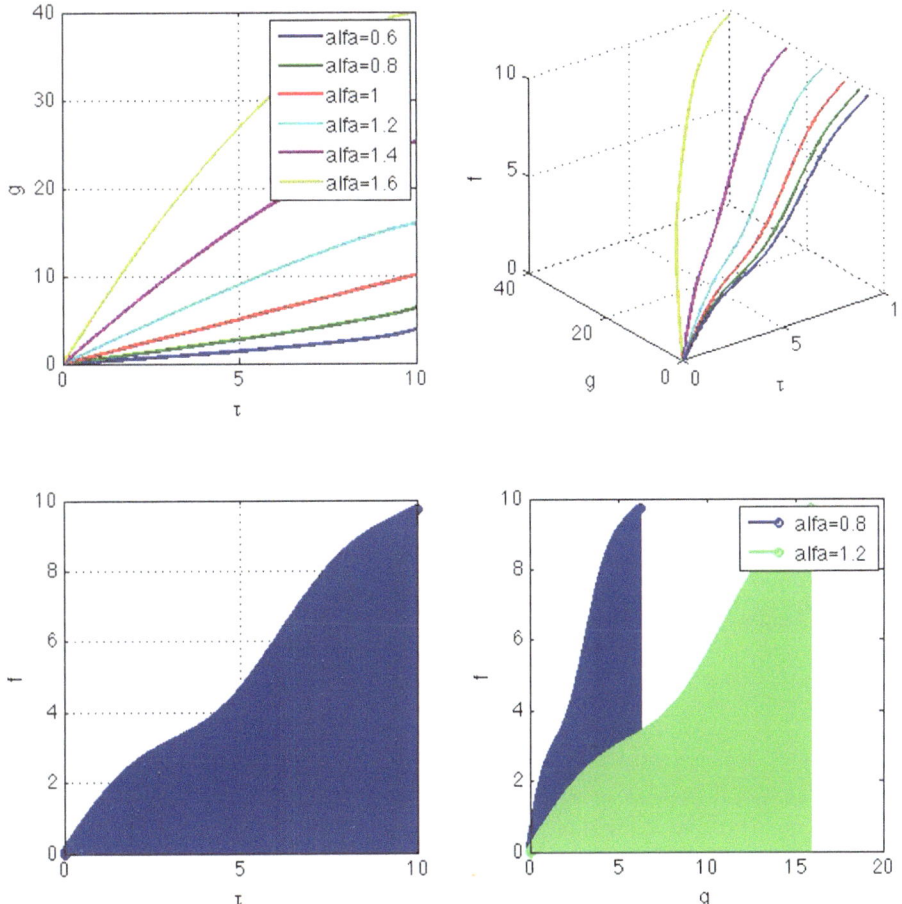

Fig. 4.6 Shadows on the wall

But this is a geometric interpretation and we, like Newton, are more interested in a physical interpretation.

Suppose that the variable τ is time and then $g(\tau)$ can be interpreted as a *warp* of the time scale. What could the meaning be? As we have seen in the very definition of the second, the measurement of time is essentially the result of counting the number of times that a recurring phenomenon occurs, but we are not able to verify if the time *absolute* that occurs between any pair of those consecutive events is exactly the same as that between any two others, giving rise to the possibility of time scale inhomogeneity. Let's imagine that these two possible scales are those represented by the variable τ and the function $g(\tau)$, that the first corresponds to how we see the seconds pass on our watch while we time ourselves on a race to know the distance traveled, and the second to how an external observer who knows

Fig. 4.7 Deformation of time

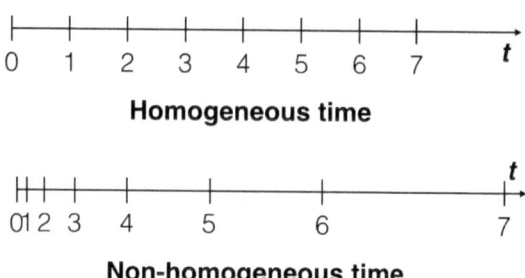

Homogeneous time

Non-homogeneous time

Table 4.1 Speed measurements

Individual Chrono (s)	Measured Speed (m/s)	Absolute Chrono (s)
0	10	0
1	11	1
2	12	3
3	13	7
4	12	15
5	11	31
6	10	63
7	9	127

that our clock does not work correctly counts them. For instance, as shown in the Fig. 4.7: our measurement is based on assuming that the clock works correctly and we measure seven units of time, while the external observer knows that the interval between each click of the second hand doubles in *duration* the previous interval. The result would be the one shown in the Table 4.1. If we do the math, we will have thought to go 79 m (Riemann integral with $f(t) = v(t)$, speed) but in reality we will have gone 1368 m (Stieltjes integral).

But let's notice that $g(\tau)$ not only depends on τ but also on t, which represents the last value measured by the individual that moves. Thus, if we change t, $g(\tau)$, our cosmic time changes. Definitely [7]:

> Each observer must have his own measure of time, which is what would be recorded by a clock moving alongside him, and identical clocks moving with different observers [in different places and with different speeds] would not have to coincide.

If we go from Physics to Biology, we again find these distorted time scales. Pierre Lecomte Du Noüy, French biophysicist, member of the Rockefeller Institute and director of the Division of Molecular Biophysics of the Pasteur Institute, published in 1937 a book entitled *Biological Time* [8]. In it, based on experimental measurements on the speed of wound healing, he obtains the curve shown in Fig. 4.8, in which it can be seen that the relative healing speed (*physiological time*) and the appreciation of time (*psychological time*) is a

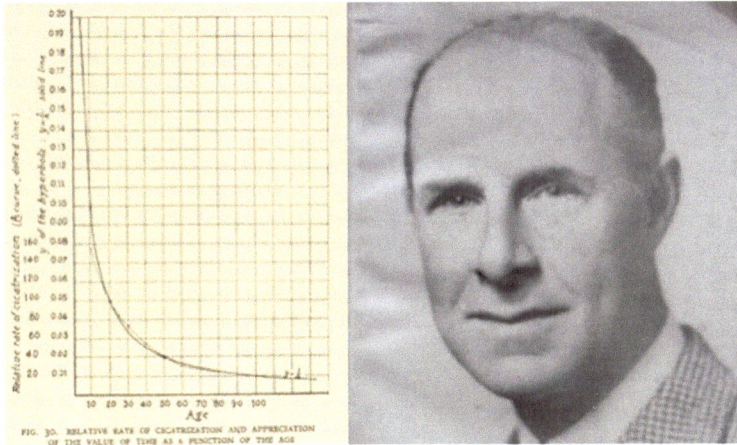

Fig. 4.8 Pierre Lecomte Du Noüy, and an illustration of his book *Biological Time*

function of age: one year for a child of ten represents the same length of time (*duration*) as three or four years for his parents of forty. That is, again, absolute time is the same for both, child and parents, but durations are different.

In August 2003, an article entitled *The strange story of Peter Lynds* [9] was published in *The Guardian*, in which the author, David Adam, wondered if this young New Zealander scientist was a Physics genius comparable to Albert Einstein following the publication of his article in *Foundations of Physics* [10] and the controversy that said publication gave rise. The core of this controversy was that the author used as new concepts that–something that apparently went unnoticed by the reviewers–had already appeared in the writings of authors such as Henri Bergson and, later, Alfred N. Whitehead.

Henri Bergson, philosopher, Nobel Prize for Literature in 1927, profoundly revised the foundations of science, our thought and our conception of time and life, placing in orbit, between others, the idea of *duration*. He spoke of a new science that, according to his own words, «... would be a *mechanics of transformation*, and our *mechanics of translation* would be a particular case of it, a projection onto the plane of pure quantity» [11].

These projections, so metaphorical, can very well be represented with the Stieltjes integral as we have seen above. For his part, Alfred North Whitehead, mathematician, philosopher, co-author with Bertrand Russell of *Principia Mathematica*, following in part in the footsteps of Bergson, wrote [12]:

> Our observational 'present' is what a call a 'duration'. It is the whole of nature apprehended in our immediate observation. It has therefore the nature of an event, but possesses a peculiar completeness which marks out such durations as a special type of events inherent in nature. A duration is not instantaneous. It is all that there is of nature with certain temporal limitations. In contradistinction to other events a duration will be called infinite and the other events are finite.

Fig. 4.9 Sampling laws

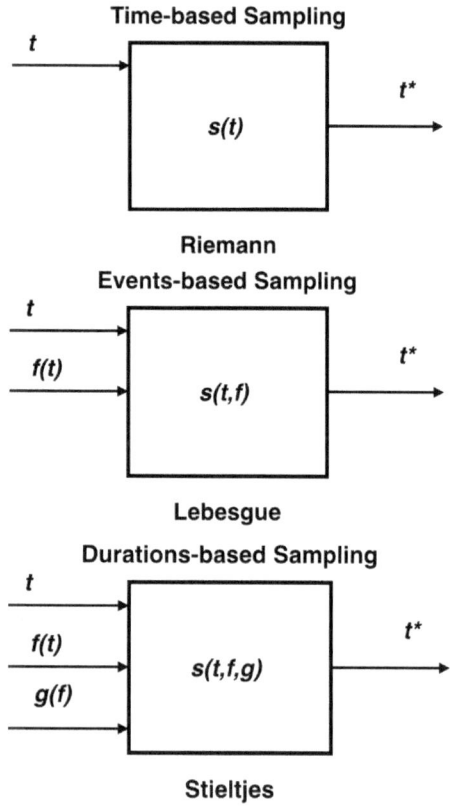

In the last decades, Control Theory has significantly expanded its field of applications. Today it is easy to find works on the control of biological systems at macro, micro and nano scales, of quantum systems, and of systems that try to emulate the functioning of the human being. As we can see, one only has to follow the path traced by Nature to suggest a *Stieltjes-type sampling* that would lead to a *duration-based control*.

Summarizing, we can establish laws according to which the sampling instants are determined as a function of:

1. A smoothly flowing time: Riemann sampling/time-based control;
2. The values or increments of the sampled function: Lebesgue sampling/event-based control;
3. The values of a function that depends on time and the sampled function: Stieltjes sampling/duration-based control.

This is represented in Fig. 4.9.

References

1. S. Dormido, J. Sánchez, E. Kofman, Muestreo, control y comunicación basados en eventos. Revista Iberoamericana de Automática y Electrónica Industrial **5**(1), 5–26 (2008)
2. K.J. Aström, B.M. Bernhardsson, Comparison of Riemann and Lebesgue sampling for first order stochastic systems, in *Proceedings of the 41st IEEE Conference on Decision and Control* (Las Vegas, Nevada, USA, 2002), pp. 2011–2016
3. T. Mann, *La montaña mágica* (Plaza & Janés, 1985)
4. E. Gluskin, Nonlinear sampling and Lebesgue's integral sums, in *2010 IEEE 26-th Convention of Electrical and Electronics Engineers in Israel* (Eilat, Israel, 2010), pp. 000736–000740
5. C.J. Ishama, K.N. Savvidou, El tiempo y la física moderna, *El tiempo*, ed. by K. Ridderbos (Cambridge University Press, 2003), pp. 11–28
6. I. Podlubny, Geometric and physical interpretation of fractional integration and fractional derivation. Technical Report TUKE-10-2001, Technical University of Kosice, Kosice, Eslovaquia, 2001
7. S.W. Hawking, *Historia del tiempo. Del big bang a los agujeros negros* (Crtica, 1988)
8. P.L. Du Noüy, *Biological Time* (Macmillan, 1937)
9. D. Adam, The strange story of Peter Lynds (2003). https://www.theguardian.com/education/2003/aug/14/research.highereducation
10. P. Lynds, Time and classical and quantum mechanics: indeterminacy versus discontinuity. Found. Phys. Lett. **16**(4), 343–355 (2001)
11. H. Bergson, *La evolucin creadora* (Espasa–Calpe, 1973)
12. A.N. Whitehead, *Concept of Nature* (Cambridge University Press, 1990)

Behold, I will bring again the shadow of the degrees, which is gone
down in the sun dial of Ahaz, ten degrees backward. So the sun
returned ten degrees, by which degrees it was gone down.

Isaiah 38:8, King James Version

5.1 The Making of the State Space

5.1.1 From the Beginning

We had left the development of the Control Theory at the time when the fundamental contributions of Ragazzini and his disciples were produced, and we have delved into sampling,
the basis for digital control, for a while. The other path was opened by Rudolf Kalman and
constitutes what has been called *Modern Theory of Control*, the one based on representation
in the *State Space*.

Between 1958 and 1963 the works published by Kalman laid the foundations of this
theory, making important contributions such as [1]:

- The *Kalman filter*, an iterative algorithm for the optimal estimation of an unknown time
  series from an observed one;
- The notions of *controllability* and *observability*;
- The *linear quadratic regulator (LQR)* and the *gaussian quadratic regulator (LQG)*, fun-
  damental in the development of optimal control;
- The theory of *Riccati equations*, differential or algebraic;
- The notion of state of a dynamic system in an input/output context;
- The *theory of realization*, as a representation of a dynamic input/output system in the
  state space form.

B. M. Vinagre, *Time in Control Theory*, Synthesis Lectures on Electrical Engineering,
https://doi.org/10.1007/978-3-031-54042-4_5

The sources, explicit or implicit, in which these works were inspired were:

- The concepts of *phase space* and *state variables* developed at the end of the 19th century by Ludwig Boltzmann, Henri Poincaré, and Willard Gibbs, as a conceptual model and representation framework;
- Claude E. Shannon's *Mathematical Theory of Communication*, as an example of the development of a 'pure theory' independent of the particular models considered and even of the methods used for description and analysis [2];
- The theory of *optimal filtering* by Norbert Wiener;
- The computer and the developments in discrete systems (*discrete time systems*) carried out by Ragazzini and his disciples Jury and Zadeh.

For mechanical systems, phase space usually consists of a two-dimensional space in which the trajectories that follow position and momentum are represented as a function of time. This would be the position of Poincaré in his *Mécanique Céleste*; but the story comes from behind.

Phase space was used originally to describe specific types of dynamical systems, but today has become synonymous with the idea of a large parameter set linked to the degrees of freedom of the considered system, like it was in the cradle.

We could start as far as with the greeks and his study of *physis* (V century BC), Plato (424/423–348/347 BC), Aristotle (384–322 BC), Lucretius (99–55 BC), Thomas Aquinas (1225–1274) or Descartes (1596–1650) following the trail of the *causality principle*; or with Newton (1642–1727) and his mechanics, where

> The motion of a system of particles is fully determined for all future time by the present positions and momenta of the particles and by the present and future forces acting on the system. How the particles actually attained their present positions and momenta is immaterial. Future forces can have no effect on what happens at present [3, p. 154].

Or Laplace (1749–1827) and his considerations on *determinism*:

> Thus we have to consider the present state of the universe as the effect of its previous state and as the cause of the one that is to follow. An intelligence that at a certain moment knew all the forces that animate nature, as well as the respective situation of the beings that compose it, if it were also broad enough to subject such data to analysis, could include in a single formula the movements of the largest bodies of the universe and those of the lightest atom; nothing would remain uncertain and both the future and the past would be present before his eyes [4, p. 25].

But, with [5], we will begin our story in the midst of XIX century.

With no reference to dynamics, nor mention of phase space, Liouville in 1838 studied a system of n first-order differential equations

$$\frac{dx_i}{dt} = P(t, x_1, x_2, \ldots, x_n), \tag{5.1}$$

with the complete set of solutions

$$x_i = x_i(t, a_1, a_2, \ldots, a_n), \tag{5.2}$$

being the a_i arbitrary constants. In 1842 Jacobi recognized that this kind of differential equations could describe mechanical systems, particularly those implying, in Hamiltonian dynamics, the position coordinates, x_i, and the momentum coordinates, p_i.

But at this moment, there was no concept of space beyond our *physical three dimensions*. The idea of a generalized space, a multidimensional one, takes long time to access the minds, even of the most conspicuous mathematicians and scientists, the time required to extend the concept of variable from algebraic entities to coordinate axes, a way that culminates with Riemann in 1868 (see [5]).

It was in Boltzmann's 1871 papers about the kinetic theory of gases, when he mades the analogy between physical trajectories of particles in a two-dimensional space and Lissajous figures, and in 1872 it uses for the first time the term *phase*. Then, in 1879, Maxwell adopted the Boltzmann's terminology to describe the *state* of a system:

> We have hitherto, in speaking of a phase of the motion of the system, supposed it to be defined by the values of the n co-ordinates and the n momenta. We shall call the phase so the defined the phase (pq). (Cited in: [5])

The final step, the putting of phase into space, was given by Poincaré at the end of the century. In 1885 the Swedish mathematician Mittag-Leffler suggested to the King Oscar II of Sweden to offer a mathematical prize in honor of his 60th birthday, being the topic the problem of finding a general solution to the stability of the solar system. Poincaré was attracted by the topic, very close to his works, and, even not solving the problem, his work was the seed for the developments included in *Les méthodes nouvelles de la mécanique céleste* (1892–1899), where he stablished the foundations of the phase space and the state variables.

5.1.2 Rudolf Kalman

For Control Theory, the final step was given by Rudolf Kalman. In [3] Kalman deals with the mathematical descriptions on dynamical systems. Though the focus is in linear systems, the study is relevant for our purposes.

He considers two models of physical systems, the one based on *vector differential equations*, i.e., phase space and state variables, and the one based on transfer functions. Two different «languages» whose use was, at that time, «surrounded by confusion». In Kalman's view [3, p. 152] the difficulty is due to insufficient appreciation of the concept of a *dynamical system*. Control theory is supposed to deal with physical systems, and not merely with mathematical objects such as a differential equation or a transfer function. We must therefore pay careful attention to relationship between physical system and their representation via differential equations, transfer functions, etc.

i.e., we must pay careful attention to relationship between reality and models, between the territory and the map. The Kalman's proposal can be read as follows:

1. To establish a precise, abstract definition of a (physical) dynamical system based on generalizations of the Newtonian causality.
2. To formulate the central problem, the problem of realization, as follows: given an (experimentally observed) impulse response matrix, how can we identify the (linear) dynamical system which generated it?
3. To give the conditions and procedures for obtaining the abstract description of the coupling between external variables (inputs and outputs) and the internal variables (states), and for being the system so described an irreductible realization of a given impulse matrix.

By translating Newton to modern terminology, we can say, with Kalman, that the numbers which specify the instantaneous position and momentum of each particle represent the *state* of the system, an abstract quantity figuring the «minimal amount of information about the past history of the system which suffices to predict the effect of the past upon the future. Further, we say that the forces acting on the particles are the *inputs* of the system. Any variable in the system which can be directly observed is an *output*» [3, p. 154].

From the point of view of Mathematics, Kalman gives an axiomatic definition of dynamical systems:

Definition 5.1 A dynamical system is a mathematical structure defined by the following axioms:

1. There is given a *state space* Σ and a set of values of *time* Θ at which the behavior of the system is defined; Σ is a topological space and Θ is an ordered topological space which is a subset of the real numbers.
2. There is given a topological space Ω of functions of time defined on Θ, which are the admissible *inputs* of the system.
3. For any initial time t_0 in Θ, any initial state x_0 in Σ, and any input u in Ω defined for $t \geq t_0$, the future states of the system are determined by the transition function φ : $\Omega \times \Theta \times \Theta \times \Sigma \rightarrow \Sigma$, which is written as $\varphi_u(t; t_0, x_0) = x_t$. This function is defined only for $t \geq t_0$.
4. Every *output* of the system is a function $\psi : \Theta \times \Sigma \rightarrow reals$.

5. The functions φ and ψ are continuos, with respect to the topologies defined for Σ, Θ and Ω and the introduced product topologies.

and *phase space* [3, p. 154–156]:

> Giving a fixed value of (t, x) is equivalent to specifying at some time, t, the state, x, of the system. We shall call (t, x) a *phase* and $\Theta \times \Sigma$ the *phase space*.

For the case of *real, finite dimensional* ($\Sigma = R^n = n$-dimensional linear space), *continuous time* ($\Theta = R^1$) and *linear* (φ is linear on $\Omega \times \Sigma$ and ψ is linear on Σ), and functions φ and ψ sufficiently "smooth", such a system is governed by the *dynamical equations*

$$\frac{dx}{dt} = F(t)x(t) + G(t)u(t), \tag{5.3}$$

$$y(t) = H(t)x(t), \tag{5.4}$$

defined on $-\infty < t < \infty$, where x, u, and y are n, m, and p-vectors respectively, and the matrices $F(t)$, $G(t)$, and $H(t)$ are continuous functions of the time t.

No less interesting for us is the following statement

> To justify our claim–implicit on the above discussion–that equations (5.3) and (5.4) are a good model of physical reality, we wish to point out that these equations can be *concretely simulated*[1] by a simple physical system: a general-purpose analog computer. Indeed, the numbers (or functions) constituting F, G, and H may be regarded as specifying the "wiring diagram" of the analog computer which simulates the system [3, p. 156].

Just change analog by digital, wiring diagram by program, and this physical reality, the system, becomes an entirely logical, artificial, mathematical entity, and the state space frame a place that can assume non-causality and no link with the "real" world, even linked to real numbers.

A step forward is done when we talk about "equivalent" dynamical systems. Here Kalman is clear again:

> The state vector x must always be regarded as an abstract quantity. By definition, it cannot be directly measured. On the other hand, the inputs and outputs of the system (5.3) and (5.4) have concrete physical meaning. Bearing this in mind, equations (5.3) and (5.4) admit two interpretations:
>
> 1. They express relations involving the abstract linear transformations $F(t)$, $G(t)$, and $H(t)$.
> 2. At any fixed time, we take an arbitrary but fixed coördinate system in the (abstract) vector space Σ. Then the symbol $x \equiv (x_1, \ldots, x_n)$ is interpreted as the numerical n-tuple consisting of the coordinates of the abstract state vector which is also denoted by x. F, G,

---

[1] The emphasis is mine.

and H are interpreted as the matrix representations of the abstract linear transformations denoted by the same letters under (1).

According to Kalman, from the point of view of Physics a dynamical system must be defined in terms of quantities which can be *directly observed*. For this, in the case of linear systems we can proceed by measuring the impulse-response matrix $S(t, \tau)$ by applying to each input, with the system at rest, an impulse (ideally, or a very narrow and sharp pulse) and observing the effects on the outputs. The question of interest here is «When and how does the impulse-response matrix determine the dynamical equations of the system?» [3, p. 159]. In other words, *when and how the dynamical equations have a physical meaning?*

If we succeed in this task, we will call to (5.3) and (5.4) a *realization* of $S(t, \tau)$. But this only means that the system can be built using the standard techniques for simulation, analog or digital computers, and have no relation with the physical reality of the system. We have only «get closer to physical reality», because still stand the question of «how much of the physical world can be determined from a given amount of experimental data» [3, p. 159]. We can summarize the main results as follows [6]:

1. Despite appealing to Physics, the state space is considered a n-dimensional real vector space X.
2. Any *basis* in X is a set of n vectors linearly independent.
3. The are infinite many basis that can generate the space X.
4. The coordinates x_i with respect to some fixed basis are called the *states variables*.
5. A point in this space is called the *state*.

When we are considering dynamical systems, and chiefly for control:

1. There are internal variables called *state variables* whose values at a fixed time $t = t_0$ form the *state*.
2. There are external variables that can drive the system called *inputs*.
3. There are external variables which can be directly observed called *outputs*.
4. The state is the smallest collection of numbers which must be known at time $t = t_0$ in order to be able to predict the outputs of the system for any time $t \geq t_0$, knowing the inputs.
5. A state space model is an abstract structure describing the coupling between the external variables and the internal variables.
6. Given an experimentally observed impulse-response matrix, we can identify a linear dynamical system which generated it. This system is called a *realization* of the impulse-response matrix.
7. Rigorously speaking, a state space model is an *irreducible* realization of the given impulse-response matrix, a realization with minimal dimension.

5.2   Irreversibility

5.2.1   Irreversibility and States

This approach to dynamical systems, the State-Space approach, whose origins and foundations, at the risk of boring the reader, I have allowed myself to recall, is a heir of Newtonian
mechanics through Poincaré's celestial mechanics, and as such it shares his concept of time:
«The music of the spheres is a palindrome, and the book of astronomy reads the same
backward and forward», because the fundamental laws of the Newtonian mechanics are not
altered by transforming the time variable t into $-t$ [7, p. 31–32]. But going down from the
celestial bodies, things can be different.

   Eduardo Sontag, a student of Kalman at the Center for Mathematical Systems Theory,
in his book *Mathematical Control Theory* [8, pp. 158–164, 181], introduces a concept in
which I now want to stop: *reversibility*. In simple words, a system is *weakly reversible* if
it can recover a previous state: state z^0 is *reachable* from state x^0, $z^0 \in \mathcal{R}(x^0)$, if and only
if state x^0 is *reachable* from state z^0, $x^0 \in \mathcal{R}(z^0)$. It is *strongly reversible* if the same path
that takes us from x^0 to z^0 can be traveled backward. Considering this concept, a system is
completely controllable only if, in addition to fulfilling the range condition, it is reversible.

   These concepts of reversibility are nicely illustrated in the Chap. 23 of *Sylvie and Bruno*
by Lewis Carroll, entitled «An Outlandish Watch». The narrator enters a little town and
encounters two fishermen's wives talking. Having a «Magic Watch» he decides to wait until
the conversation is over to do an experiment with it: he turned the clock back a minute and
saw that the two women and their conversation were back to the starting point. Just as the
sundial of Ahaz went ten degrees backward. Wanting to explore other possibilities of the
magic watch, he soon finds the opportunity when he witnesses an accident in which a young
man riding a bicycle collides with a box that had fallen from a cart onto the street. And then,
moving back the hand of the watch, he saws, «almost without surprise this time, all things
restored to the places they had occupied at the critical moment when I had first noticed the
fallen packing-case». But, unfortunately, although he had taken care to remove the box and
watched with satisfaction as the bicycle passed and disappeared without a hitch, he foresees
«as the hand of the Watch touched the mark, the spring-cart [...] was back again at the door,
and in the act of starting, while [...] the wounded youth was once more reclining on the heap
of pillows, his pale face set rigidly in the hard lines that told of pain resolutely endured».
Outwitted by the watch, he tried its reverse action. When he passes in front of a small house
with a garden and sees the door open, he decides to burst into a quiet family scene, four
girls with their mother, seated by the fire and doing some needlework, and observes how
the entire course of events–labor, actions, conversation–were reversed by the action of the
watch. The two first experiments can be considered, if we neglected the final results, as cases
of weak reversibility, while the last one is of strong reversibility.

   But in real world final results can not be neglected. Suppose we want to prepare a rice
paella with monkfish, prawns, clams and prawns. Let's start with the ingredients: 400 g of

rice, 200 g of diced clean monkfish, 200 g of peeled prawns, 200 g of clams, 8 prawns, fish broth, salt, parsley, 1 finely chopped onion, 1 finely chopped carrot, 1 green bell pepper finely chopped, 1 finely chopped tomato, 2 garlic cloves finely chopped. Once the ingredients are properly prepared, the preparation of the paella comes: (1) in the paella, poach or sauté the vegetables for 5 min; (2) when it is well poached, add the fish, the prawns, and the clams; (3) fry well and add the rice; (4) move it and add the broth; (5) taste for salt and when it starts to boil, put the prawns on top and cook for 15 min over low heat until done.

Even in this little detailed recipe, we can see that it is a process in which, in order to obtain a certain amount of final product of a certain quality, we need determined amounts of suitably prepared constituent products, and adding them in a dosed manner in various stages or threads. And we can see that the process is not reversible, and that the quality of the final product depends on the quality of the constituent products. This is what usually happens in *batch processes*.

Going to more technical examples, we can cite the one studied in [9] about the penicillin production process. The control objective is to maximize penicillin production using the feed rate, u, with a fixed substrate concentration, S_{in} as a control variable. The state-space model of the process is:

$$
\begin{bmatrix} \dot{x}_1 \\ \dot{x}_2 \\ \dot{x}_3 \\ \dot{x}_4 \end{bmatrix} = \begin{bmatrix} \mu(x_2)x_1 \\ -\frac{\mu(x_2)x_1}{Y_1} - \frac{\rho x_1}{Y_2} \\ \rho x_1 \\ 0 \end{bmatrix} + \begin{bmatrix} -\frac{x_1}{x_4} \\ \frac{1}{x_4}(S_{in} - x_2) \\ -\frac{x_3}{x_4} \\ 1 \end{bmatrix} u \tag{5.5}
$$

$$
\mu(x_2) = \mu_{max} \frac{x_2}{K_m + x_2 + x_2^2/K_i} \tag{5.6}
$$

being x_1, x_2, x_3 the concentrations of substrate, biomass, and penicillin, x_4 the volume of the reactor, μ_{max}, K_m, K_i, ρ, kinetic parameters, and Y_1, Y_2 parameters associated with the efficiency of the reaction. Studying these equations we can see that the variables x_3 and x_4 are irreversible, *as any variable that is always increasing.*

Because they are *irreversible*, they are chronologically tied to their initial conditions, and suffer:

- Loss of controllability;
- Increase in the effects of uncertainties;
- Increase in the effects of disturbances;
- Limitation of the control action;
- Strong dependence on initial conditions.

5.2.2   Irreversibility and Equilibrium

In 1948, Norbert Wiener, mathematician, philosopher, professor at MIT, author of fundamental contributions in electronic engineering, communications and control systems among other fields, published a book extraordinary entitled *Cybernetics or Control and Communication in the Animal and the Machine*, coining and defining the term at the same time. The Chap. 1 begins, based on a popular German song, with the comparison between astronomy and meteorology and the radical difference that exists in the quality of the predictions of both sciences. The Babylonians could already predict an eclipse, and their prediction was so accurate that they used this type of phenomenon to measure time. But not even today we can predict the weather a week from now with the same accuracy. I quote [7, p. 32]:

> Finally, when all this [astronomy, celestial mechanics] was reduced by Newton to a formal set of postulates and a closed mechanics, the fundamental laws of this mechanics were unaltered by the transformation of the time variable t in its negative.

> Thus if we were to take a motion picture of the planets, speeded up to show a perceptible picture of activity, and were to run the film backward, it would still be a possible picture of planets conforming to the Newtonian mechanics. On the other hand, if we were to take a motion-picture photograph of the turbulence of the clouds in a thunderhead and reverse it, it would look altogether wrong. We should see downdrafts where we expect updrafts, turbulence growing coarser in texture, lightning preceding instead of following the changes of cloud which usually precede it, and so on indefinitely.

> What is the difference [...] which brings about all these differences, and in particular the difference between the apparent reversibility of astronomical time and the apparent irreversibility of meteorological time?

Relative to the space in which they are found, celestial bodies constitute a relatively small number of particles of different sizes in an otherwise empty space, punctual if we consider the distance that separates them, rigid. All of this makes it easy to compute their position and momentum, and maintain past-future symmetry. Nothing is like this in the case of meteorology, and nothing is like this in any system made up of a space relatively full of particles of similar sizes, not rigid, and between which there are large coupling forces. In these cases it is impossible to compute position and momentum of the *microstates* but only probability distributions for *macrostates* (temperature, pressure, volume, etc.), and the time is directional (there is the *arrow of time*). These phenomena, like almost all biological ones, escape the Newtonian idea of time and approach the Bergsonian one, unidirectional, creative in its advance: a process of cell division or morphogenesis is more like the making of a paella than the movement of a connecting rod.

Reversibility works for Newtonian mechanics based systems and processes, but not for many everyday ones. Many processes are irreversible, abandoning the Newtonian time and entering a realms were the *time's arrow* exist. Batch processes, discrete-event systems [10], biological processes, or thermodynamic processes, could lost reversibility. When are irreversible, in them the phenomena occur in a certain order, time is *directional*. They all share an essential characteristic: they evolve *far from equilibrium* [11, 12]:

Reversible systems are those that evolve close to a state of equilibrium, ... and it seems that time
will not pass through them, while in systems that evolve far from equilibrium, the directionality
of time is evident and for this reason they are called systems irreversible.

And this characteristic connects irreversibility to steam machine, thermodynamics, and
entropy [13]. And with entropy to *Information Theory* [14, Chap. 2, The Origins of Information Theory]:

By this [reversible process] we mean that if work is done in pushing the piston slowly back
against the gas and so recompressing it to its original volume, the exact original energy pressure,
and temperature will be restored to the gas. In such a reversible process, the entropy of the
gas remains constant, while its energy changes. Thus, entropy is an indicator of reversibility;
when there is no change of entropy, the process is reversible.

5.3   Reversibility and Tiny Things

This same characteristic is found in the so-called *molecular motors*, biological machines
essential in the movement of living organisms on a molecular scale [15]. A distinctive
element of these motors with respect to their macroscopic counterparts is that they operate
in what could be called a thermal bath, an environment in which fluctuations due to noise
are of great importance. Due to the stochastic models used to study them, they are also
known as *Brownian motors*. These motors rectify thermal fluctuations to obtain directional
movement using spatial or temporal asymmetry. An illustration of this phenomenon is given
by the so-called *Feyman-Smoluchowski ratchet* (Fig. 5.1).

The device consists of a ratchet (a toothed wheel that rotates freely in one direction but not
in the opposite due to the trigger) connected by a shaft to a bladed wheel immersed in a fluid
at temperature T_1. The molecules of the fluid constitute a thermal bath and are subjected to
Brownian motion with an average kinetic energy that depends on the temperature. The device
is supposed to be small enough that the momentum from a single molecular collision can spin
the blades. Although such collisions would tend to rotate the shaft in both directions with

Fig. 5.1 Feyman-
Smoluchowski ratchet
(*source* [15])

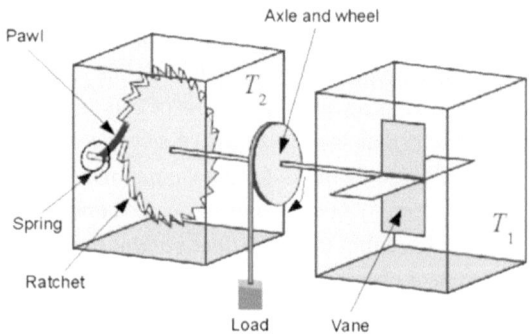

equal probability, the ratchet only allows it to rotate in one. The unidirectional movement of the device can be used to do some work, such as lifting a weight against gravity.

We can obtain a simplified mathematical model of such devices using the equation

$$m\ddot{x}(t) = -V'(x(t)) - \gamma\dot{x}(t) + \xi(t),\tag{5.7}$$

where:

- $V(x)$: ratchet potential, periodic of period L ($V(x) = V(x + L)$) and spatially asymmetric ($V(x) = V_0(\sin(kx) + A\sin(nkx)$, $k = 2\pi/L$) (Fig. 5.2),
- $\xi(t)$: white noise (thermal),
- $\gamma\dot{x}(t)$: force due to friction.

At these scales, the inertia term is usually negligible, giving rise to the *Langevin equation*:

$$\gamma\dot{x}(t) = -V'(x(t)) + \xi(t).\tag{5.8}$$

In Figs. 5.3 and 5.4 we can see the block diagrams corresponding to Eqs. (5.7) and (5.8), and different trajectories of a particle driven by brownian motion.

If at the *nanoscale* we found irreversibility, at the *microscale* we have to deal with reversibility. One of the main challenges of *microrobotics* for medical applications (which «allows the doctor to situate himself inside the patient» [16, pp. 58–62]) is to perform a *nonreciprocal* (irreversible) motion to move within the organism. And it runs into a problem

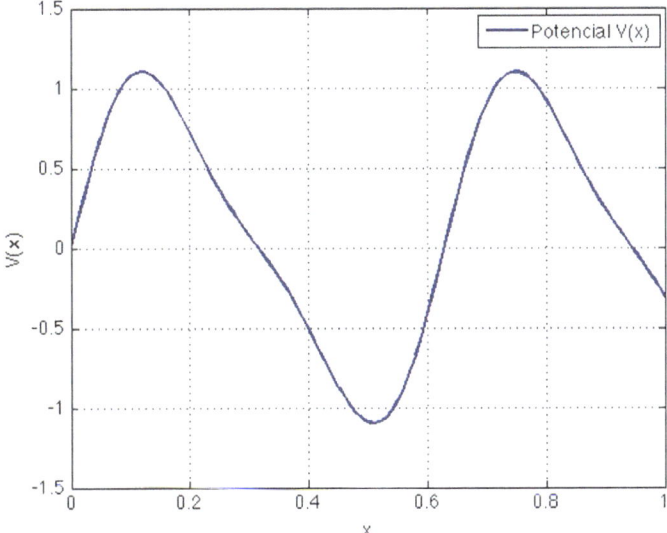

Fig. 5.2 Example of ratchet potential

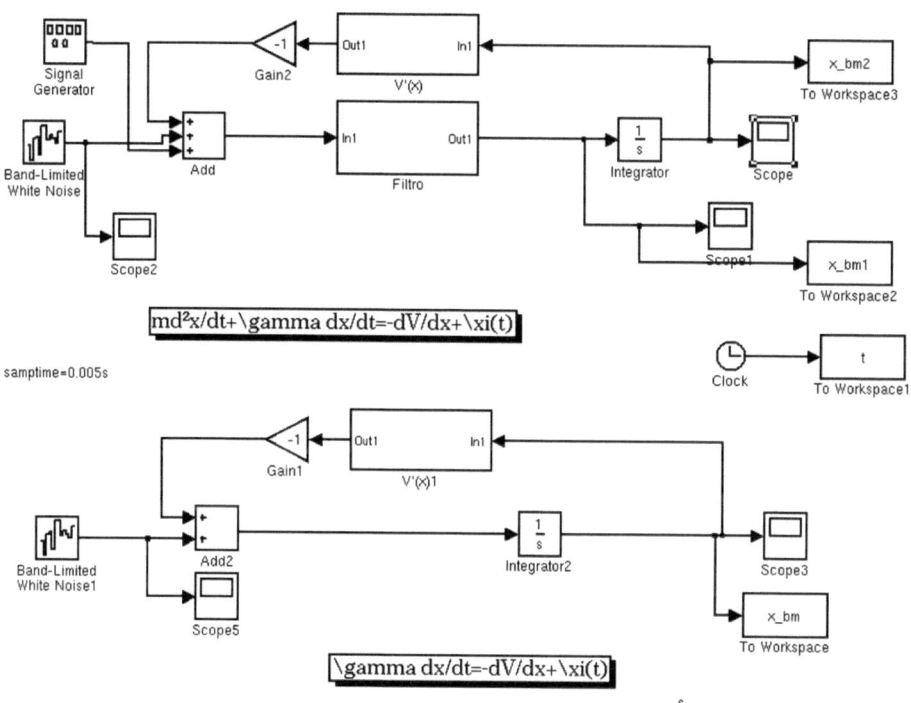

Fig. 5.3 Block diagrams of the brownian motors corresponding to (5.7) and (5.8)

due, precisely, to reversibility: inertial forces, as we have seen, are no longer the most important, ceding their preponderant role to others such as viscous forces.

Robots, due to their size, have the need to navigate in a *Stokes or laminar flow*. This regime is imposed by the low values of the Reynolds number,[2] and gives rise to a perfect reversibility in time, quite the opposite of what Wiener commented on meteorology and thermodynamics: there is no arrow of time. As E. M. Purcell stated «Time, in fact, makes no difference–only configuration» [17]. This makes macroscopic propulsion mechanisms, which use reciprocal motion (like that of a stiff oar) ineffective: due to the importance of viscous forces, there is no slip that takes advantage of the inertial forces to move. Then, we have to emulate some kind of ratchet potential throught the gait, and mimic, again, nature and move like bacteria move (see Fig. 5.5).

---

[2] A common measure of the ratio of the inertial to viscous forces

$$R_e = \frac{\rho l v}{\eta},\qquad(5.9)$$

where ρ is the fluid density, v is the velocity of the object, l is a characteristic length of the object, and η is the fluid viscosity.

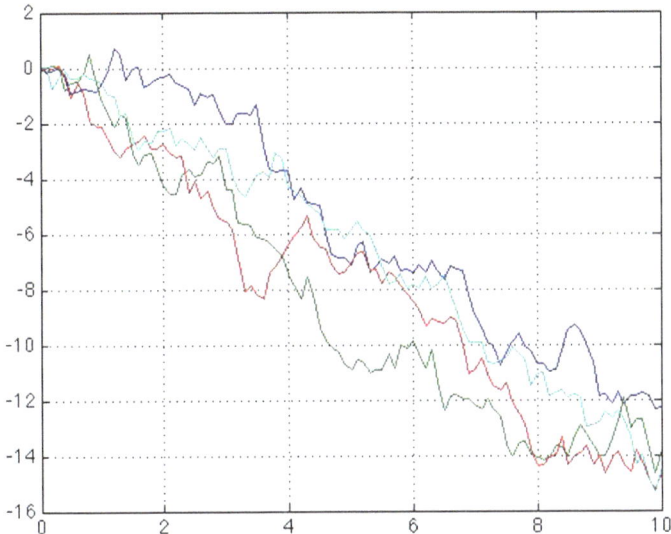

Fig. 5.4 Different trajectories of a particle driven by brownian motion

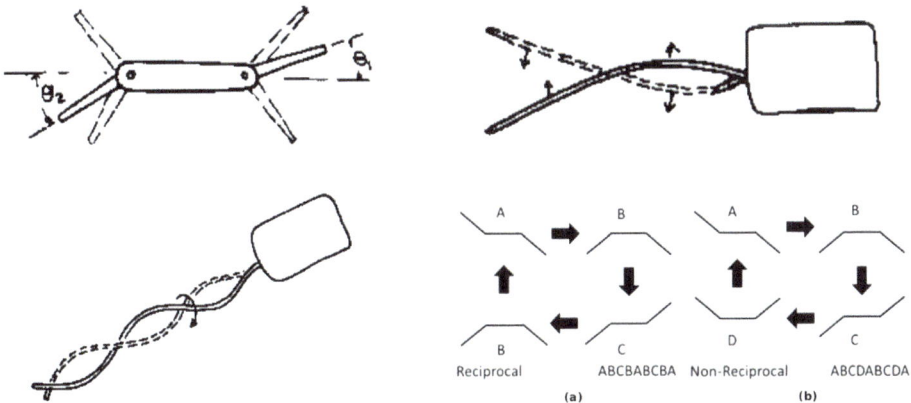

Fig. 5.5 The Purcell swimmer, forms of motion (flexible oar and corkscrew), and reciprocal/nonreciprocal motion scheme. (Taken from [17])

References

1. J.C. Willems, Laudatio for the Johann Bernouilli lecture for Rudolf Kálmán (1991). http://www.math.rug.nl/bernoulli/vorigelezingen/lezing01/laudatio01.pdf
2. R. Kálmán, On the general theory of control systems, in *Proceedings of the 1st IFAC Congress on Automatic Control*, vol. 1 (Mosce, URSS, 1961), pp. 481–492

3. R.E. Kalman, Mathematical description of linear dynamical systems. J.S.I.A.M. Control A **1**, 152–191 (1963)
4. P.-S. de Laplace, *Ensayo filosófico sobre las probabilidades* (Alianza, 1985)
5. D.D. Nolte, The tangled tale of phase space. Phys. Today. April, 33–38 (2010)
6. R.E. Kalman, On the general theory of control systems, in *Proceedings of 1st International Congress on Automatic Control*, (1960), pp. 481–492
7. N. Wiener, *Cybernetics: or Control and Communication in the Animal and the Machine* (The M.I.T. Press, 1985)
8. E. Sontag, *Mathematical Control Theory* (Springer, 1998)
9. L.M. Gómez, H. Botero, H. Álvarez, F. di Sciascio, Análisis de la controlabilidad de estado de sistemas irreversibles mediante teoría de conjuntos. Revista Iberoamericana de Automática e Informática Industrial **12**, 145–153 (2015)
10. B. Hrúz, M.C. Zhou, *Modeling and Control of Discrete-event Dynamic Systems* (Springer, 2007)
11. W.M. Haddad, V.S. Chellaboina, S.G. Nersesov, Time-reversal symmetry, Poincaré recurrence, irreversibility, and the entropic arrow of time: From mechanics to system thermodynamics. Nonlinear Anal. R. World Appl. **9**, 250–271 (2008)
12. Lina María Gómez and Hernán Álvarez, La irreversibilidad: una mirada desde la teoría de sistemas de control. Revista Avances en Sistemas e Informática **8**(2), 31–39 (2011)
13. M. King, *Process Control: a Practical Approach* (John Wiley & Sons, 2011)
14. J.R. Pierce, *An Introduction to Information Theory* (Signals & Noise. Dover Publications Inc, Symbols, 1980)
15. P. Reimann, Brownian motors: noisy transport far from equilibrium. Phys. Rep. **361**, 57–265 (2002)
16. P. Ball, Feynman's fancy. Chem. World. 58–62 (2009)
17. E.M. Purcell, Life at low Reynolds number. Am. J. Phys. **45**, 3–11 (1977)

6.1   PID Controller and the Sequence of the Time

> *It's a poor sort of memory that only works backwards.*
>
> Lewis Carroll, *Through the Looking-Glass*, Chap. V, «Wool and Water».

If we continue reading the *Confessions* by Agustine of Hippo, we encounter the following statement [1, XI, XX, 26]:

> But what is now clear and manifest is that there are neither past nor futures, nor can it be properly said that the times are three: past, present and future; but perhaps it would be more proper to say that there are three times: present of past things, present of present things, and present of future things. [...] present of past things (memory), present of present things (vision) and present of future things (expectation).

In this quotation are implicit, on the one hand, a linear conception of time in which all Western culture is immersed, including of course its science and technology, and on the other hand the idea that the present, the moment to act, is what is relevant, the point where past and future converge.

In former chapters we have considered two views of time. In the one hand, as Heraclitus said, time is but a succession of present moments, and only change exists (panta rhei–everything flows). On the other hand, we can think with Parmenides that time is duration and change is only an illusion. But may be, «we must attempt to understand time as a correlation of future, past and present [...] in which the self-conscious individual proves to be the instrument...» [2, p. 57].

© The Author(s), under exclusive license to Springer Nature Switzerland AG 2024   63
B. M. Vinagre, *Time in Control Theory*, Synthesis Lectures on Electrical Engineering,
https://doi.org/10.1007/978-3-031-54042-4_6

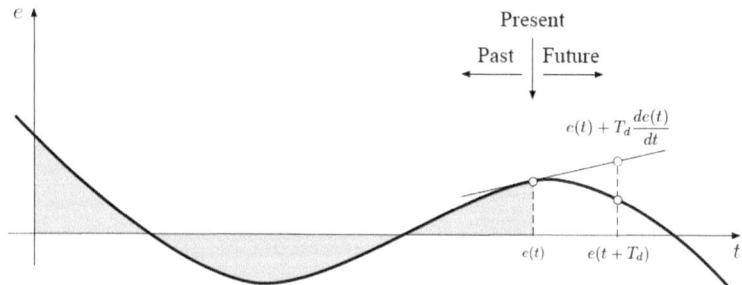

Fig. 6.1 Time along real axis and PID actions over error signal (from [3])

And can't these words be understood as a metaphorical description of an algorithm well familiar to all of us, the PID (see Fig. 6.1)? One the one hand, it acts according to our vision (observation, measurement), but not only taking into account its current value (present–proportional), but also the residue of its past history (memory–integral) and the expectation of its future evolution (prediction–derivative). On the other hand, the controller is the «self-conscious individual» that performs the correlation of past, present and future.

Maybe the beauty of the proportional-integral-derivative (PID) algorithm for feedback control, its simplicity and efficiency have more deeper roots. But let's leave metaphysics.

The proportional-integral-derivative (PID) controller is recognized as the most common form of feedback. In process control today, more than 95% of the control loops are of PID type, but these controllers can be found in all areas where control is used. Despite its simplicity, or perhaps because of it, PID controllers have survived many changes in technology, from mechanics and pneumatics to microprocessors via electronic tubes, transistors, or integrated circuits, among others. Actually, practically all PID controllers made today are based on microprocessors, so this electronic element has had a dramatic influence on this kind of control providing PIDs additional advances features, such as gain scheduling, continuous adaptation, and automatic tuning [4].

Any corrective action in control requires true understanding of nature of error, which represents how far or near is the desired output from actual output. This is basis of any control algorithm. But more information can be obtained about the error looking at way it is shaping up in the past and its variation. Thus, proportional, integration and derivation actions applied to error can help designers to characterize the nature of the error and, accordingly, adjust the control variable.

The key aspect when tuning PID controllers is in deciding how to best combine those three terms or modes to achieve the most efficient regulation of the process variable for the considered problem. As well known, the most obvious way is to use a simple weighted sum where each term is multiplied by a tuning constant or gain, and the results then are added together as follows:

$$u(t) = K_p \left(e(t) + \frac{1}{T_i} \int_0^t e(\tau)d\tau + T_d \frac{de(t)}{dt} \right), \tag{6.1}$$

where $u(t)$ is the control signal, $e(t)$ is the control error ($e = y_{sp} - y$, i.e., the difference between the desired set-point, y_{sp}, and the measured process variable, y), and K_p, T_i, and T_d are the controller parameters: proportional gain, integral time constant, and derivative time constant, respectively.

Control law (6.1) guarantees that the present (due to the proportional action), the past (by means of the integral action) and the future of the error (by the derivative action) are taken into account, as shown in Fig. 6.1. Two main observations can be made to Eq. (6.1): (1) the controller needs only compute the current error between the measured process variable and the desired set-point to calculate how much and how fast that difference has been changing over time, and (2) the relative contributions of each term then can be adjusted by choosing appropriate values of the controller parameters.

Equation (6.1) corresponds to the standard algorithm or ideal form of PID controller. It can be also written as

$$u(t) = K_p e(t) + K_i \int_0^t e(\tau)d\tau + K_d \frac{de(t)}{dt}, \tag{6.2}$$

where $K_i = K_p/T_i$ and $K_d = K_p T_d$ are the integral and derivative gains, respectively. This form is sometimes known as theoretical algorithm because these gains, together with K_p, are the weighting factors for the proportional, integral, and derivative terms.

However, unlike parameters T_i and T_d, K_i and K_d have no physical meaning.

6.1.1   Going into Detail About Parameters

Proportional action

The proportional action decreases the deviation between the set-point value y_{sp} and the system output y (i.e., the error) with the increasing of proportional gain K_p. A pure proportional controller can be given by:

$$u(t) = K_p e(t) + u_b, \tag{6.3}$$

where u_b is a polarization or reset. Indeed, when $e = 0$, the control signal reduces to $u(t) = u_b$. The parameter u_b usually takes the value $(u_{max} + u_{min})/2$, where u_{max} and u_{min} are the maximum and minimum limits of the actuator, respectively; however, sometimes u_b can be set manually to a value that ensures that the steady-state error is zero at a given set-point.

Likewise, the proportional gain in some PID controllers is specified in terms of its inverse, known as the proportional band (P_b). More precisely, P_b represents the percent change in the error signal necessary to cause a full-scale change in the proportional action. As can be observed in Fig. 6.2, the tendency of y to oscillate increases as P_b decreases. The large oscillations occurring with a small P_b are due to the fact that the power is reduced very quickly when the system output enters the proportional band, meaning a balanced state cannot be established immediately. Indeed, confusing "proportional band" with "proportional gain"

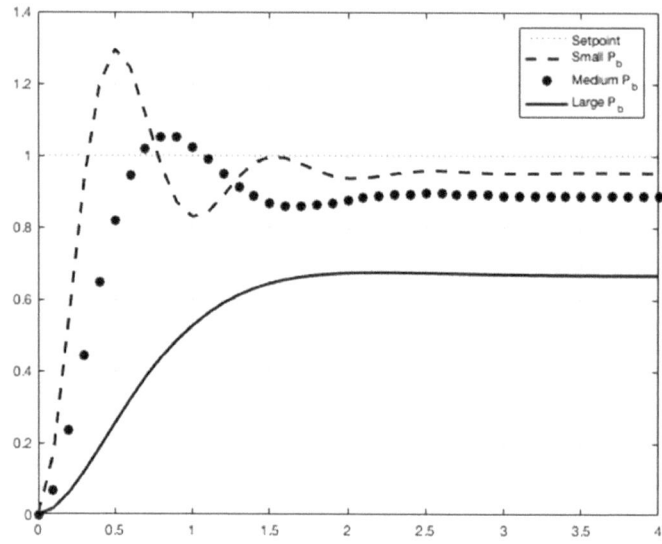

Fig. 6.2 System response for different P_b

leads to a decreased proportional action when the control engineer wants more, and vice-versa.

Integral action

The main function of the integral action is to guarantee that the steady-state error is zero, i.e., the output of the controlled system is equal to the desired set-point in steady-state. The following simple explanation proves this affirmation. Let consider a system controlled by PID controller (6.1) in steady-state with both the control signal (u_{ss}) and the error (e_{ss}) being constant. If this is the case, from (6.1) it can be stated that the control signal will be given by

$$u_{ss} = K_p \left(e_{ss} + \frac{e_{ss}}{T_i} t \right) \tag{6.4}$$

While $e_{ss} \neq 0$, this clearly contradicts the hypothesis that the control signal u_{ss} is constant. Integral action was known originally as a device with automatic reset because the integral term seemed to automatically adjust the set-point to the exact value required to eliminate the steady-state error caused by the proportional term. Figure 6.3 shows the classical implementation scheme of a PI controller with automatic reset. From the block diagram, the control signal can be obtained as

$$u = K_p e + I, \tag{6.5}$$

with

Fig. 6.3 PI controller scheme (classical implementation with automatic reset)

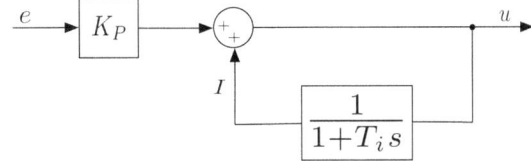

$$I = U(s)\frac{1}{1 + T_i s} \Rightarrow I + T_i \frac{dI}{dt} = u. \tag{6.6}$$

Substituting (6.6) in (6.5), the control signal is given by

$$u = K_p \left(e(t) + \frac{1}{T_i} \int e(\tau)d\tau \right). \tag{6.7}$$

Derivative action

The objective of the derivative action is to improve the system stability in closed-loop. Derivative controllers give responses to changing error signals but do not, however, respond to constant error signals, since with a constant error the rate of change of error with time is zero. Because of this, the derivative term is combined with, at least, the proportional term.

The derivative action of a PD controller can be viewed as a crude prediction of the error in future on the tangent line of the error curve at time t, being T_d the prediction horizon (see Fig. 6.1). (Actually, the derivative action uses linear extrapolation, not prediction.) The basic structure of the controller is

$$u(t) = K_p \left(e(t) + T_d \frac{de(t)}{dt} \right) \approx e(t + T_d). \tag{6.8}$$

That is, the control signal is proportional to an estimation of the error at time T_d forward over a straight line tangent to the error curve. But different prediction horizons (in fact, linear and non linear extrapolations) for the error can be obtained by choosing other methods of approximation (see [3]).

The classical implementation of a fractional derivative action is illustrated in Fig. 6.4. From this figure, the following relation can be obtained

Fig. 6.4 Derivative action scheme (classical implementation)

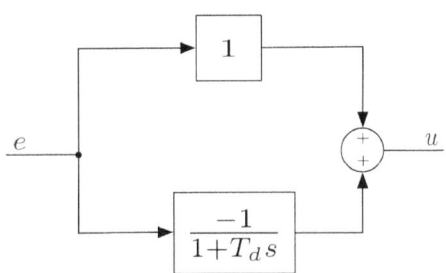

$$U(s) = \left(1 - \frac{1}{1 + T_d s}\right) E(s) = \frac{T_d s^\mu}{1 + T_d s^\mu} E(s), \tag{6.9}$$

which corresponds to a derivative action with noise filter. It is well known that the derivative part of the PID controller requires low-pass filtering to limit the high-frequency gain.

6.2   Feedback: Mixing Memory and Desire

> April is the cruellest month, breeding
> Lilacs out of the dead land, mixing
> Memory and desire...
>
> T. S. Eliot, *The Waste Land.*

Consider the system described by the discrete convolution sum

$$y(k) = \sum_{-\infty}^{k} u(n), \tag{6.10}$$

where $u(.)$ is the input sequence and $y(.)$ the output sequence. It is clear that for obtaining the output we need to perform more and more computations (sums) as time flows, i.e., as the temporal index k grows. The direct application of Eq. (6.10) characterises a system without memory, a system for which the time passage run without a trace; a system that does not learn from the experience, from the events of the past.

Fortunately, in this case we can solve the problem of obtaining the accumulated sum of sequential inputs in an easier and computationally more efficient way by rearranging the previous expression in the form

$$y(k) = \sum_{-\infty}^{k-1} u(n) + u(k) = \underbrace{y(k-1)}_{} + u(k), \tag{6.11}$$

where the term marked $y(k-1)$ is nothing more than the last sum obtained. Thus, saving this previous sum in each iteration, we only have to add the current input to obtain the total sum up to the present moment. We can graphically represent this equation using the block diagram in Fig. 6.5. In this diagram we can identify one loop which brings the output to the input after delaying it by one sample. So, using this scheme to perform the computations is more efficient in terms of computational complexity than the one of crude sum. At the same time it is more elegant, more intelligent, because it uses *memory* learning from past events: it uses *feedback*.

Let's agree that any system processes (input–output) energy or information. Then, we can say that, in a broad sense, feedback denotes that some of the output energy (information) is returned as input [5]. This is the characteristic advantage of using feedback: it is a way

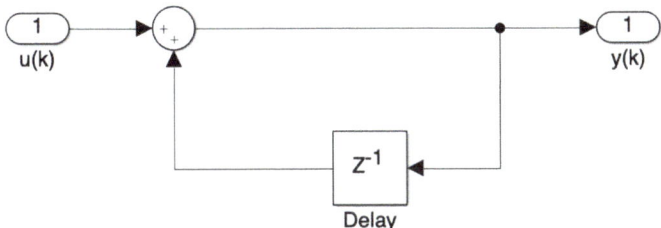

Fig. 6.5 Block diagram of the discrete accumulator corresponding to Eq. (6.11)

to learn from the past (instinct or experience) for achieving a goal, like in animal taxis that involves stimulus-induced movement towards a specific direction, such as moths flying towards a lightbulb. And because the lessons learned come from the past and the goal is fixed in future, feedback connects *memory and desire*. In fact, so important is this advantage, that feedback is a universal principle or mechanism allowing not only the construction of self-regulating machines, but the existence of life itself: the homeostasis or the ecosystems equilibrium are nothing but the natural paradigms of the machines that we design and build.

In the particular case at the beginning of this section, the discrete accumulator, the feedback is *positive*: the fraction of the output that contributes to the input has the same sign as the original input, and so adds to it. In general, the effect of positive feedback is to reinforce a behavior, like the naive example of the accumulator. But this behavior uses to be a lot more complex and interesting: positive feedback is involved in all processes showing hysteresis or memory, in cell differentiation and in immunology [6], as well as in childbirth: always the result will be a building process that culminates, after surpassing a threshold, in a physiological event in time and the return of the system to a new equilibrium (Fig. 6.6) [7].

On the other hand, in *negative* feedback the signals from the output are used, by decreasing the effective input, to restrict the output in order to achieve the objective and remain within an established margin around it. This is the feedback modality used for engineers when designing self-regulating machines, and the one involved in physiological homeostatic regulation, the body control of physiological variables (temperature, blood pressure, ion and

Fig. 6.6 Evolution of output in positive feedback culminating with an event and the return to equilibrium or stasis (adapted from [7])

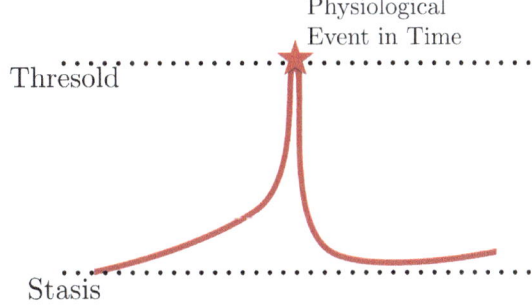

mineral concentrations, etc.) within precise and optimal ranges where life may be sustained. In these cases, it is required that some signals from the goal be used for modifying the activity of the system «in the course of the behavior» [5].

In engineering, positive feedback is used as a principle for designing memory storage units in digital electronics or autonomous oscillators, to generate bistable switches and clocks, while negative feedback enables robust regulation [8]. But in general, both life and the higher level of technology require the use of *cycles, loops*, with both positive and negative feedback and also feedforward, a characteristic denoted in the cerebral circuits as *recurrence* or *circularity* [9, 10]. Combining positive and negative circuits, past and future to act in the present, feedback is the simplest and more efficient form of memory, be implemented in flips-flops or in our brains; and memory, storing time, is an essential ingredient of consciousness and a key for building our identity.

6.3   Mind and Time: Control for IA

> The days of thinking of time as a river
> — evenly flowing, always advancing —
> are over.
>
> David Eagleman, «Brain time».

In the first quarter of the twenty first century, the era announced by Samuel Butler in the 1870's [11], that of the intelligent machines that will succeed man in the supremacy of the Earth, seems to be close. But there are two aspects, at least, that we have to solve to give birth our successors, and both are related with time: how to achieve the smooth and subtle processing of our brain, and how to produce in the machine the *duration*.

The neuroscientist David Eagleman opens the work cited in the epigraph with the following story [12]:

> At some point, the Mongol military leader Kublai Khan (1215–94) realized that his empire had grown so vast that he would never be able to see what it contained. To remedy this, he commissioned emissaries to travel to the empire's distant reaches and convey back news of what he owned. Since his messengers returned with information from different distances and traveled at different rates (depending on weather, conflicts, and their fitness), the messages arrived at different times. Although no historians have addressed this issue, I imagine that the Great Khan was constantly forced to solve the same problem a human brain has to solve: what events in the empire occurred in which order?

Our brain is connected with the world by means of sensory information via the nervous system. But these informations come, like Kublai Khan's emissaries, from different distances traveling at different speeds; and, furthermore, their messages are processed at different speeds in different parts of the brain. This is a hard control problem: multirate, asynchronous and distributed, with the additional difficulty on the processor of multitasking. The final result is the building of «temporal illusions».

Roughly speaking, the brain solves the multirate and multipath problem by following a recursive procedure at different levels.[1] First it uses different clocks for each sense, and each clock opens temporal windows of different widths waiting for the arrival of the slowest information concerning an object. By doing so, the brain can perform *feature-binding*, that is, to keep an object's features perceptually unite. Second, it proceeds in the same way (to wait to collect all information), and constantly recalibrates its expectations about arrival times in order to obtain *temporal-binding*, the assignment of the correct timing of events in the world. This is an ingenious combination of synchronous and asynchronous sampling, adaptive, time-driven and event-driven control that allows us to interpret the world and apprehend the causality.

Though subjective, these problems are still clearly linked to the hardware, to the body. But how to manage with those subjective expansions and contractions of time that seems to be linked to the *mind*? We have to learn how time and memory are linked, how age varies our duration-based sampling weighting function; and, finally, how to implement all this in a machine. For this last step, two ways affecting control theory are open nowadays: *neuromorphic control* and *morphological computation*.

Neuromorphic control is based on neuromorphic electronic systems developed by Carver Mead three decades ago. Taking inspiration from biological systems, he realized that in them information is represented by the relative time of arrival of nerve spikes that are digital in amplitude but analogue (asynchronous) in time [13, 14]. Used for the first time in [15], it has been a kind of Sleeping Beauty for control community until very recent times, when the *intelligence* based on bits inherited from Alan Turing seems to be reaching its limits, and we turn our heads towards animal intelligence based on spikes, action potentials [8, 16]. As illustrates Fig. 6.7, the stimulus enters the neuron and, reached a given threshold, the spikes are triggered as a physiological even that paced the time.

On the other hand, the morphological approach proposes to delegate part of the functionality (detection, computation and control) to body morphology [17]. The fundamentals of this approach comes from both, neurophysiology and robotics [18–20], and can be included in the set of theories about the extended, embodied or distributed mind. I want just to remark

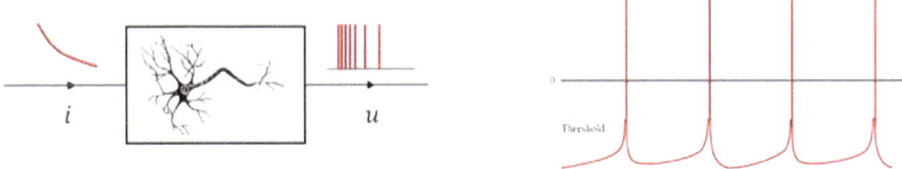

Fig. 6.7 A neuron and the triggered spikes (adapted from [16])

[1] See [12] for a more precise and detailed information.

here that, being this a step forward in the way of the new *imitation game*, for including the body we will have to manage the same problems that the brain does (different paths and different speeds for information arrival from the sensors, different structures and different rates for information processing and different ways for information reaching the effectors) to provide our machines with the capabilities of feature and temporal binding, so that they can build their own *temporal illusions*.

References

1. S. Agustín, *Confesiones* (Bruguera, 1984)
2. J.T. Fraser, *The Voices of Time* (George Braziller, 1966)
3. I. Tejado, B.M. Vinagre, J.E. Traver, J. Prieto-Arranz, C. Nuevo-Gallardo, Back to basics: Meaning of the parameters of fractional order PID controllers. Mathematics 7(530), 2 of 16 (2019)
4. K.J. Aström, R.M. Murray, *Feedback Systems: an Introduction for Scientists and Engineers*, 2nd edn. (Princeton University Press, 2008)
5. A. Rosenbleuth, N. Wiener, J. Bigelow, Behavior, purpose and teleology. Philos. Sci. **10**(1), 18–24 (1943)
6. J. Demongeot, M. Kaufman, R. Thomas, Positive feedback circuits and memory. C.R. Acad. Sci. Paris, Sciences de la vie/Life Sciences **323**, 69–79 (2000)
7. Noel Ways. Feedback systems: an introduction to negative and positive feedback systems with emphasis on homeostasis and stress. https://noelways.com
8. L. Ribar, R. Sepulchre, Designing multiscale mixed-feedback systems. IEEE Control. Syst. (December), 34–63 (2021)
9. A.J. Bell, Levels and loops: the future of artificial intelligence and neuroscience. Philos. Trans. R. Soc. Lond. B **354**, 2013–2020 (1999)
10. D. Eagleman, *Incógnito: las vidas secretas del cerebro* (Anagrama, 2011)
11. S. Butler, *Erewhon o Allende las montañas* (Bruguera, 1982)
12. D.M. Eagleman, Brain time, in *What's Next? Dispatches on the Future of Science: Original Essays from a New Generation of Scientists*, ed. by M. Brockman (Vintage, 2009), pp. 155–169
13. C. Mead, Neuromorphic electronic systems. Proc. IEEE **78**(10), 1629–1636 (1990)
14. C. Mead, How we created neuromorphic engineering. Nat. Electron. **3**, 434–435 (2020)
15. S.P. DeWeerth, L. Nielsen, C.A. Mead, K.J. Åström, A simple neuron servo. IEEE Trans. Neural Netw. **2**(2), 248–251 (1991)
16. A. Serrano, B.M. Vinagre, I. Tejado, Fractional neuromorphic controller for a servomotor: analog implementation. IFAC PapersOnLine **56**(2), 4301–4306 (2023)
17. H. Hauser, T. Nanayakkara, F. Forni, Leveraging morphological computation for controlling soft robots. IEEE Control. Syst. (June), 114–129 (2023)
18. R. Brooks, Intelligence without representation. Artif. Intell. **47**, 139–159 (1991)
19. A. Clark, *Supersizing the Mind: Embodiment, Action, and Cognitive Extension* (Oxford University Press, 2008)
20. T. Fuchs, The circularity of the embodied mind. Front. Psychol. **11**(August):13, 1707 (2020)

Coda

In 1958, the psychologists Robert H. Knapp and John T. Garbutt published an article where, incardinated in an experimental inquiry into variations in time judgment and attitudes as a function of personality, they concluded that the subjects under study could be classified into three clusters [1]: the Dynamic-Hasty Cluster, the Naturalistic-Passive Cluster, and the Humanistic Cluster. The first one is «related directly to achievement motivation, and we suspect to that body of thought and ideology which has been described under the term "Protestant ethic". In a sense, it reflects a Newtonian sense of time, one defined in terms of an absolute, impersonal, constant, and directional rate of change in the universe.» The second one «appears to us quite different in its ideological home. Essentially it is an oriental or mystical time sense [...] Conspicuous here is the lack of any suggestion of directionality and the prevailing sense of time as surrounding and encompassing in a passive sense, almost as though time were an oceanic medium.» Finally, in the third one, «it seems implicit that "man is the measure of all things," as the great Sophist, Protagoras, long ago put it. Such a view might be identified with classical Mediterranean thought.» The particular experiments considered in the study involved a group of 73 male undergraduate students of a university in Connecticut, which can be considered a very biased universe. Furthermore, 1958 is quite far from us. But there is an aspect of the study that, in my opinion, is still interesting: that personality traits correlate with time imagery and with attitude towards time. Because those traits are the basis of individual choices and, by that, of societal orientations: they form the *Weltanschauung*[1] of individuals, societies and ages, as well as of the different fields of the human knowledge and activities. Until now we mainly were Dynamic-Hasty, like Newton in Fig. 1: portrayed as a muscular youth, sitting on a rock, crouching over a diagram, and measuring it with a compass. In it the British poet and illustrator William Blake satirises Newton's scientific approach to the world for being too reductive: he is so fixated on his

---

[1] Cosmovision, the way of conceiving and interpreting the World.

B. M. Vinagre, *Time in Control Theory*, Synthesis Lectures on Electrical Engineering, https://doi.org/10.1007/978-3-031-54042-4

Fig. 1 Newton, William
Blake's monotype (1795–1805)
at the Tate Britain

calculations that he is blind to the world around him. Blake wrote in a letter from November 1802 to his friend Thomas Butts [2]:

> Now I a fourfold vision see,
> And a fourfold vision is given to me;
> Tis fourfold in my supreme delight,
> And threefold in soft Beulah's night,
> And twofold Always. May God us keep
> From Single vision & Newton's sleep!

The Newtonian concept of time, with such formidable consequences, seems to be reaching its limits. Disciplines such as quantum mechanics, thermodynamics, artificial intelligence or biology, seem to demand, if we want to consider them as fields of application of the theory of control, that we abandon the «Single vision & Newton's sleep»: tiny machines seem to live in «an oceanic medium» and pertain to the Naturalistic-Passive Cluster, event and duration based sampling and control approach the men as «measure of all things» that characterizes the Humanistic Cluster. We have to enter the Wonderland, ... and there things are otherwise [3, Chap. VII]:

> Alice sighed wearily. "I think you might do something better with the time," she said, "than waste it in asking riddles that have no answers."
>
> "If you knew Time as well as I do," said the Hatter, "you wouldn't talk about wasting it. It's him."
>
> "I don't know what you mean," said Alice.
>
> "Of course you don't!" the Hatter said, tossing his head contemptuously. "I dare say you never even spoke to Time!"
>
> "Perhaps not," Alice cautiously replied: "but I know I have to beat time when I learn music."

Fig. 2 Alice at the mad tea party with the Hatter, the Dormouse, and the March Hare (John Tenniel's illustration)

"Ah! that accounts for it," said the Hatter. "He won't stand beating. Now, if you only kept on good terms with him, he'd do almost anything you liked with the clock. For instance, suppose it were nine o'clock in the morning, just time to begin lessons: you'd only have to whisper a hint to Time, and round goes the clock in a twinkling! Half-past one, time for dinner!"

In Control Theory we have learned to «beat time» when using it, but if we want to continue being in «good terms with him», we must consider his many faces (Fig. 2).

References

1. R.H. Knapp, J.T. Garbutt, Time imagery and the achievement motive. J. Pers. **26**, 426–434 (1958)
2. W. Blake, *Blake's Poetry and Designs* (Norton, 1979)
3. L. Carroll, *Alice's Adventures in Wonderland* (The Project Gutenberg, 1991)

Appendix: Fractional Calculus and Time

> In the preconscious process of converting the primary data of our
> experience step by step into structures, information is necessarily
> lost, because the creation of structures, or the recognition of
> patterns, is nothing else than the selective destruction of
> information.
>
> Gunter Stent, "Prematurity and uniqueness in scientific discovery".

A.1   Abel and the Tautochrone

In Chap. 2, regarding the contributions of Huygens and Bernoulli in the development of the pendulum clock, we studied the problem of the tautochrone, the curve for which the time it takes for an object to slide along it without friction with constant gravity to the lowest point is independent of the initial position. We mention there that the first successful demonstration that the tautochrone curve is a cycloid dates back to 1659 by Christiaan Huygens in his work *Horologium Oscillatorium*. But the person who demonstrated that the cycloid is the only curve that meets the defining property of the tautochrone was Niels Henrik Abel [1]. Among the many contributions he made to mathematics in his short life (1802–1829), here we are interested in the works that contributed to generalising and solving the tautochrone problem in a more elegant way in 1823 and 1826. Furthermore, «in solving the generalisation of the tautochrone problem, Niels Henrik Abel had also developed a complete framework of what is now called the fractional calculus, or differentiation and integration of arbitrary real order» [2, 3]. In what follows, we will expose the problem in modern terminology and notation (see, e.g., [4]). For an historical exposition, see [3].

Consider the Fig. A.1, being m the mass of a particle that starting from rest at the point (x, y) slides over the curve $y = y(x)$, without friction and being its weight the only acting force, towards the origin. The arc s is the length of the curve measured from the origin to the point (u, v).

B. M. Vinagre, *Time in Control Theory*, Synthesis Lectures on Electrical Engineering,
https://doi.org/10.1007/978-3-031-54042-4

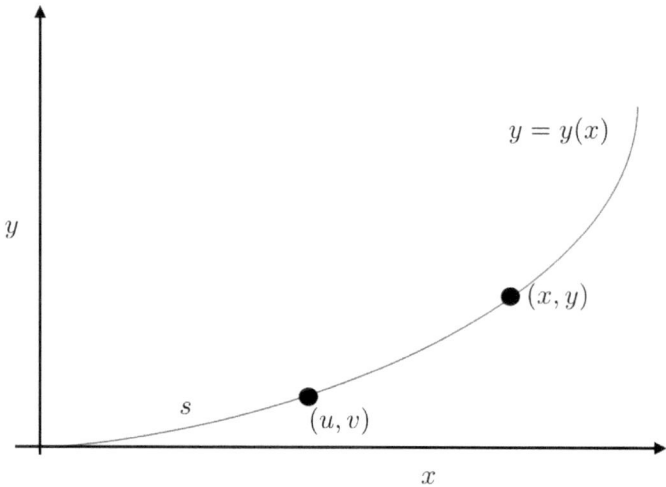

Fig. A.1 The Abel's mechanical problem and the tautochrone

In this situation we can pose two problems: (1) if the curve $y(x)$ is given, calculate the time to go from (x, y) to the origin $(0, 0)$; (2) if the time is known, determine the curve $y(x)$. Consider first the first problem.

By applying the conservation of energy principle, here only related to mechanical energy, potential and kinetic, we can state the relation

$$mgy = mgv + \frac{1}{2}m \left(\frac{ds}{dt}\right)^2,$$ (A.1)

from which we can obtain

$$\frac{ds}{dt} = \pm\sqrt{2g(y - v)}.$$ (A.2)

The arc s decreases as t increases ($\frac{ds}{dt} < 0$), so we have to choose the negative sign in Eq. (A.1). Then

$$T(y) = -\int_{v=y}^{v=0} dt = \int_{v=0}^{v=y} \frac{ds}{\sqrt{2g(y - v)}} = \frac{1}{\sqrt{2g}} \int_0^y \frac{s'(v)dv}{\sqrt{y - v}}.$$ (A.3)

After obtaining this equation, Abel continues [3]

Instead of solving this equation, I will show how one can derive s from a more general equation

$$T(y) = \int_0^y \frac{s'(v)dv}{(y - v)^n},$$ (A.4)

where n has to be less than 1 to prevent the infinite integral between two limits; $T(y)$ is an arbitrary function that is not infinite, when y equals to zero.

But in the Abel's equation (A.4) we can identify the *Caputo Fractional Derivative* or real order n, $0 < n < 1$. The solution of this equation is the inverse operation, the *Fractional Order Integral* of order n.

The most used definitions of these operators are

$$I_c^n f(t) \triangleq \frac{1}{\Gamma(n)} \int_c^t (t - \tau)^{n-1} f(\tau) d\tau, \quad t > c, \quad n \in \mathbb{R}^+, \quad (\text{A.5})$$

for the *Riemann-Liouville fractional integral* , and

$$_c D^\alpha f(t) \triangleq \frac{1}{\Gamma(m - n)} \int_0^t \frac{f^{(m)}(\tau)}{(t - \tau)^{n-m+1}} d\tau, \quad (\text{A.6})$$

for the Caputo fractional derivative, where $m - 1 < n < m$, $m \in \mathbb{N}$, and $\Gamma(.)$ stands for the Euler gamma function, an extension of the factorial function to complex numbers. We can see the Eq. (A.5) is a natural consequence of the Cauchy's formula for repeated integrals, which reduces the computation of the primitive corresponding to the n-fold integral of a function $f(t)$ to a simple convolution integral.

The Abel's mechanical problem (the second problem posed) implies that $T(y)$ is constant, and then the curve $y(x)$ can be obtained from Eq. (A.3). This can be solved by using Laplace transform and observing that Eq. (A.3) is the convolution of the functions $f(y) = s'(y)$ and $g(y) = 1/\sqrt{y}$. The solution is, of course, the cycloid.

It is worth to recall that, as we mention in Chap. 3, the possibility of derivatives of fractional order was suggested by L'Hôpital to Leibniz on 1695: «What if n be $1/2$», asked L'Hôpital; and Leibniz replied «... This is an apparent paradox from which, one day, useful consequences will be drawn». For the control community, those useful consequences gave rise to the theories of *Fractional Order Control* [5].

A.2   Fractional Integral and Duration-Based Sampling

If we consider the (left-hand sided) Riemann-Liouville fractional integral of order n (see [2])

$$I_0^n f(t) = \frac{1}{\Gamma(n)} \int_0^t (t - \tau)^{n-1} f(\tau) d\tau, \quad (\text{A.7})$$

we can write it in the form

$$I_0^n f(t) = \int_0^t f(\tau) dg_t(\tau), \quad (\text{A.8})$$

with

$$g_t(\tau) = \frac{1}{\Gamma(n+1)}\{t^n - (t-\tau)^n\}. \tag{A.9}$$

Equation (A.8) corresponds to the Riemann-Stieltjes integral considered in Chap. 4 as the basis of the Stieltjes-type sampling or duration-based sampling.

A.3   On States, Pseudo-States and Physical Quantities

By following the arguments exposed in Chap. 5, we can extend the use of Space-space control.

A.3.1   From Physics

Suppose we perform an experiment obtaining the step response represented in Fig. A.2 in blue. This could be the result of an experiment where we have observed some kind of anomalous relaxation phenomena.[2]

Then, it will be approximated by a sum of two exponential functions, i.e.,

$$y_1(t) = ae^{bt} + ce^{dt} \tag{A.10}$$

resulting the curve in red color, with $y_1(t)$: $a = 0.3059$, $b = 0.001699$, $c = -0.1654$, $d = -1.789$.

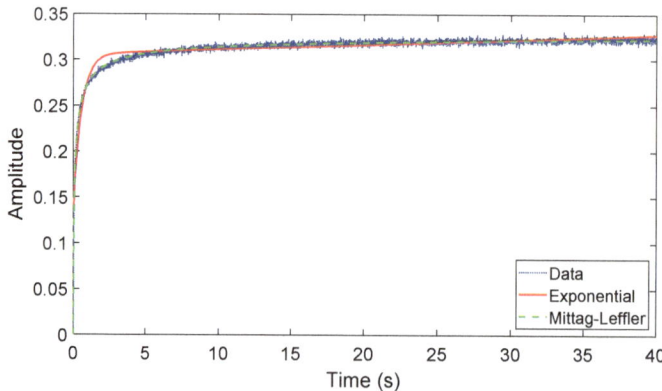

Fig. A.2 Step response of an anomalous relaxation dynamic system: fitting results

---

[2] Following Kalman's procedure, we start from the system step response. It was obtained in MATLAB by means of the command `fotf` [5] and added a noise to be more realistic.

This first fitting, the usual one, leads to an impulse-response matrix composed by two exponential terms and, consequently, to a state–space description with two state variables.

But we can try an impulse-response matrix composed by terms of the form

$$t^{\alpha-1}\mathcal{E}_{\alpha,\alpha}(-bt^{\alpha}) \qquad (A.11)$$

where $\mathcal{E}_{\alpha,\alpha}(-bt^{\alpha})$ is the Mittag-Leffler function defined as

$$\mathcal{E}_{\alpha,\alpha}(z) = \sum_{k=0}^{\infty} \frac{1}{\Gamma(\alpha(k+1))} \frac{z^k}{k!}. \qquad (A.12)$$

resulting the curve in green color with $y_2(t)$: $\alpha = 0.5449$, $\beta = 3.0284$. Then, the transfer-function matrix will be composed by terms of the form

$$Z(s) = \frac{1}{s^{\alpha} + b} \qquad (A.13)$$

and following the procedure developed by Kalman, we can obtain the dynamical equations

$$\frac{d^{\alpha}x}{dt} = F(t)x(t) + G(t)u(t), \qquad (A.14)$$

$$y(t) = H(t)x(t). \qquad (A.15)$$

where $\frac{d^{\alpha}x}{dt}$ is the Riemann-Liouville fractional derivative.[3] Now we have only one... What?, *state or pseudo-state?*

Because we have measured the impulse-response matrix $S(t, \tau)$ by applying to each input, with the system at rest, a very narrow and sharp pulse, and have observed the effects on the outputs, Eqs. (A.14)–(A.15) are a realisation of $S(t, \tau)$. In fact, we can built the system with the standard techniques of simulation. The question is: *is there any reason for considering the first model, the one based on exponentials, more 'realistic', with more 'physical meaning'?*.

All physical quantities, base or derived, «provide a rational basis for describing and analyzing the physical world in quantitative terms». They are properties of physical things or events, and they are not themselves physical things or events. And for the base quantities only the operations of comparison, addition, subtraction and multiplication by a pure number less or greater than zero, are defined in physical terms. We can clearly include in this category physical quantities as length or gravitational masses, not so clear is the case of volumes, speeds or forces. Nothing tangible is «the cube of a length, say, or the natural logarithm of a time. Products, ratios, powers, and exponential and other functions such as trigonometric functions and logarithmic are defined for *numbers*, but we have no physical correspondence in operations involving actual physical quantities.». If we take volume, or speeds, or forces

---

[3] It is worth to observe that we have used in the above few lines the three big names mentioned in Chap. 5 when talking about the origins of the state space concepts: Liouville, Riemann and Mittag-Leffler.

as base quantities is because we can define acceptable procedures to perform with them the aforementioned operations of comparison, addition, subtraction and multiplication by a pure number less or greater than zero. In fact, «the set of base quantities is very much a matter of choice», once we have defined an acceptable procedure to operate with them [6].

Something similar can be concluded for the state variables, as exposed by Kalman (see Chap. 5): we can chose any fractional derivative of, let's say, space with respect to time, to form the state vector, once we have defined the procedure for obtaining a minimal realisation of the observed physical object. From our experiment the same conclusions that G. W. Scott Blair exposes regarding his own in rheology could be derived:: «A long series of experiments [...] led us to the conclusion that firmness was judged neither by strain (γ) nor by rate of change of strain ($d\gamma/dt$) but by some entity intermediate between the two; i.e. a fractional differential. [...] When we write "$d\gamma/dt$", the "t" refers to the ordinary time of physics: "Newtonian time"» [7, 87]. But we have seen that there are a lot more 'times'.

A.3.2 From Mathematics

The idea behind this method is illustrated in Fig. A.3. Basically, the advantage of state space based control methods is due to the use of more information about the system to be controlled compared to input-output (transfer-function) based methods. System augmentation, or in general fractional derivatives, can be seen as having additional information available for control.

In this point the fractional order augmented system will be introduced. In order to introduce it let us assume the traditional integer order state-space system, given by the following equations:

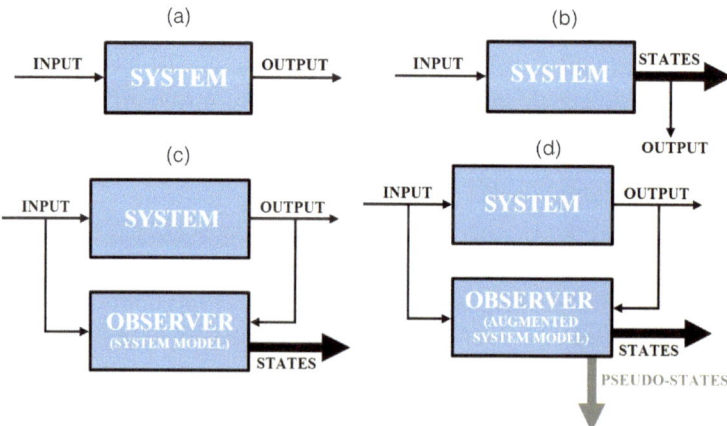

Fig. A.3 Block diagrams of: **a** Input-output system; **b** Input-output system with states; **c** Conventional observer; **d** Observer based on the augmented system

Fig. A.4 Scheme of integer order state-space system

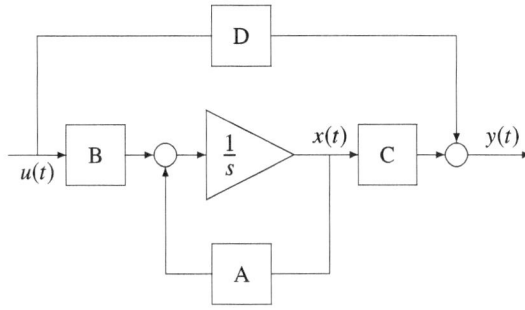

$$\dot{x}(t) = Ax(t) + Bu(t) \tag{A.16}$$
$$y(t) = Cx(t) + Du(t) \tag{A.17}$$

with initial conditions $x(0) = x_0$ and where $A \in \mathbb{R}^{n \times n}$, $B \in \mathbb{R}^{n \times m}$, $C \in \mathbb{R}^{r \times n}$. That system has the scheme of Fig. A.4.

The first order integral can be rewritten (augmented), without any additional assumptions, as a set of sequential fractional order integrals with order equal to $\frac{1}{p}$.

This leads to the system in Fig. A.5. We can state the following Theorem:

Theorem A.1 *The integer order state-space system can be rewritten in the form of fractional (rational) order system as*

Fig. A.5 Augmented fractional order system scheme

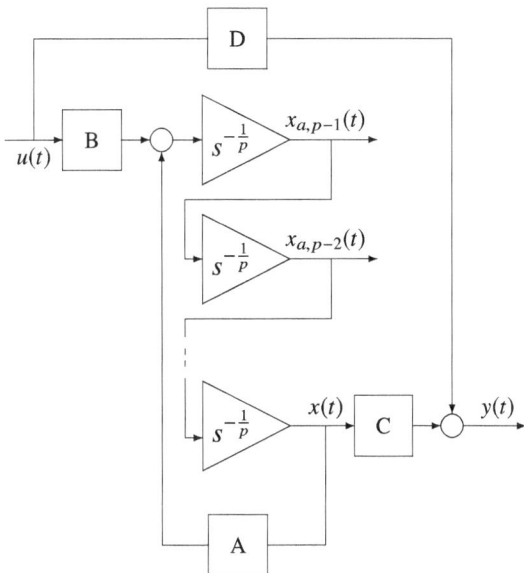

$$\begin{array}{C}C\\0\end{array}D_t^{\frac{1}{p}} \boldsymbol{x}(t) = \boldsymbol{A}\boldsymbol{x}(t) + \boldsymbol{B}u(t) \tag{A.18}$$

$$y(t) = \boldsymbol{C}\boldsymbol{x}(t) + Du(t) \tag{A.19}$$

where $\boldsymbol{A} \in \mathbb{R}^{pn \times pn}$, $\boldsymbol{B} \in \mathbb{R}^{pn \times m}$, $\boldsymbol{C} \in \mathbb{R}^{r \times pn}$ *and*

$$\boldsymbol{A} = \begin{bmatrix} 0 & I & 0 & \dots & 0 & 0 \\ 0 & 0 & I & \dots & 0 & 0 \\ \vdots & \vdots & \vdots & & \vdots & \vdots \\ 0 & 0 & 0 & \dots & 0 & I \\ A & 0 & 0 & \dots & 0 & 0 \end{bmatrix}, \boldsymbol{B} = \begin{bmatrix} 0 \\ 0 \\ \vdots \\ 0 \\ B \end{bmatrix} \tag{A.20}$$

$$\boldsymbol{C} = \begin{bmatrix} C & 0 & 0 & \dots & 0 & 0 \end{bmatrix} \tag{A.21}$$

$$\boldsymbol{x}(t) = \begin{bmatrix} x(t) \\ x_{a,1}(t) \\ \vdots \\ x_{a,p-2}(t) \\ x_{a,p-1}(t) \end{bmatrix} \tag{A.22}$$

being $p \in \mathbb{Z}^+$, *and*

$$x_{a,1} = \begin{array}{C}C\\0\end{array}D_t^{\frac{1}{p}} x(t) \tag{A.23}$$

$$x_{a,i} = \begin{array}{C}C\\0\end{array}D_t^{\frac{1}{p}} x_{a,i-1}(t) \quad \text{for} \quad i = 2 \dots p - 1 \tag{A.24}$$

the components of the augmented state vector with initial conditions

$$x(0) = \begin{bmatrix} x_0^T & x_{a,1}^T(0) & \dots & x_{a,p-1}^T(0) \end{bmatrix}^T \tag{A.25}$$

This augmentation gives the possibility to extend the estimation and control algorithms. For such a redefined system we can design a fractional order observer to estimate not only unmeasured original state variables but also unknown fractional order derivatives of those state variables. This gives us the possibility of using them for augmented state feedback control, and achieve another possibility to adjust the desired system dynamics, dynamics which is not possible to achieve by using only integer order systems (just like for the fractional PID regulators).

It is also worth to notice that we introduce into the system initial conditions which are fractional order derivatives of the trajectory $x(t)$. It is also easy to check that the solution of the system with initial conditions given as $x(0) = \begin{bmatrix} x_0^T & 0 & \dots & 0 \end{bmatrix}^T$ is exactly solution of the integer order system for the same initial conditions.

Remark A.1 It is important to bear in mind that:

1. The physical system remains at it was, because the augmented system procedure only affects to an *artificial mathematical object*, the observer.
2. In terms of vector spaces, we have increased the dimension of the complete system, like in the traditional control strategy using observers (or integral control).

The results presented by Theorem A.1 can be also applied for integer order non-linear systems given by the following equations:

$$\dot{x}(t) = f(x(t), u(t)) \tag{A.26}$$

$$y(t) = h(x(t), u(t)) \tag{A.27}$$

Theorem A.2 *The integer order non-linear state-space system can be rewritten in the form of fractional (rational) order system as*

$$
{}_0^C D_t^{\frac{1}{p}}
\begin{bmatrix}
x(t) \\
x_{a,1}(t) \\
\vdots \\
x_{a,p-2}(t) \\
x_{a,p-1}(t)
\end{bmatrix}
=
\begin{bmatrix}
x_{a,1}(t) \\
x_{a,2}(t) \\
\vdots \\
x_{a,p-1}(t) \\
f(x(t), u(t))
\end{bmatrix}
\tag{A.28}
$$

$$y(t) = h(x(t), u(t)) \tag{A.29}$$

where $p \in \mathbb{Z}^+$

A.3.3 Application Example

Let us consider the velocity of a servo system in state space form is given by:

$$\dot{x}(t) = -2.381x(t) + 26.2041u(t) \tag{A.30}$$

$$y(t) = x(t) \tag{A.31}$$

where $y(t)$ is a rotation speed of the servo (in rpm), $u(t)$ is a control signal (in Volts). The augmented fractional order system equations for the case of $p = 2$ are

$$
{}_0^C D_t^{0.5} x(t) =
\begin{bmatrix}
0 & 1 \\
-2.381 & 0
\end{bmatrix}
x(t) +
\begin{bmatrix}
0 \\
26.2041
\end{bmatrix}
u(t)
\tag{A.32}
$$

$$y(t) = \begin{bmatrix} 1 & 0 \end{bmatrix} x(t) \tag{A.33}$$

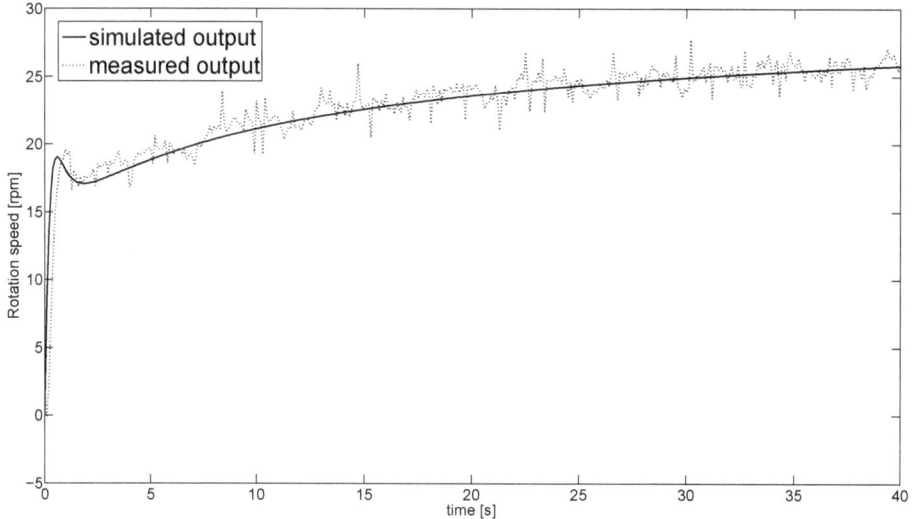

Fig. A.6 State feedback control for servo augmented fractional order system (output $y(t)$)

The cost function is chosen as

$$J_s = \int_0^{T_f} x(t)^2 u^2(t) dt \tag{A.34}$$

In this case the controller matrix is $K = [0.9146 \quad 0.2621]$

As can be observed in Figs. A.6, A.7 and A.8 the experimental results are in agreement with the simulations, and the estimated half derivatives have less noise than the direct computed ones.

Despite the many discussions about the meaning of fractional derivatives and their corresponding initial conditions, notwithstanding the appeal to the physical meaning of state variables, for control engineering purposes the state is a vector, the state space is a vector space, and no physical meaning is required as a necessary condition for its use and definitions. Furthermore, to consider as significant only the derivatives of integer order is to condemn many systems, their modeling and their control, to a limited bed of Procrustes.

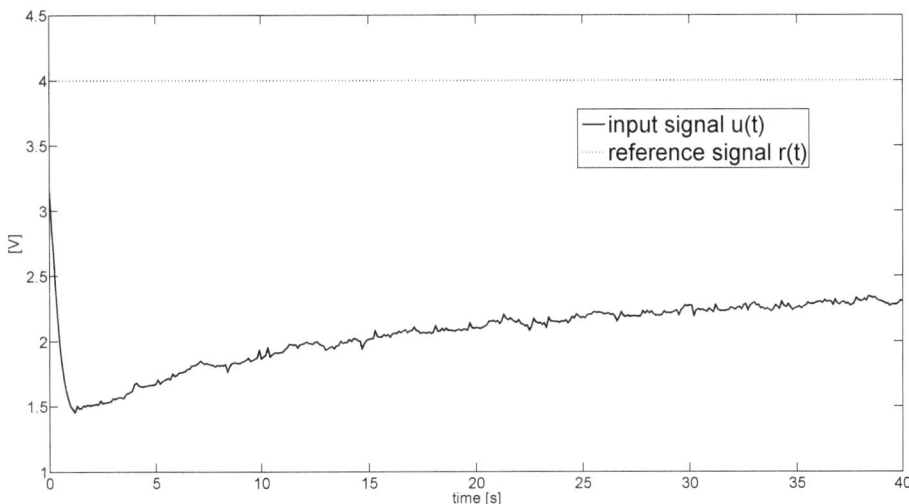

Fig. A.7 State feedback control for servo augmented fractional order system (control signal $u(t)$).eps

Fig. A.8 Direct and estimated
half order derivatives of output

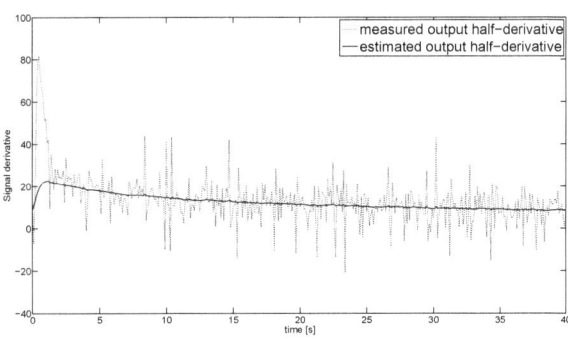

References

1. L. Yepes, La curva tautócrona: mismo tiempo de caída (2022).. https://ingenieriabasica.es/la-curva-tautocrona-mismo-tiempo/
2. I. Podlubny, Fractional-order systems and $PI^\lambda D^\mu$-controllers. IEEE Trans. Autom. Control **44**(1), 208–214 (1999)
3. I. Podlubny, R.L. Magin, I. Trymorush, Niels Henrik Abel and the birth of fractional calculus. Fract. Calc. Appl. Anal. **20**(5), 1068–1075 (2017)
4. G. Aguilar, El problema mecánico de Abel. Una aplicación sencilla del Teorema de Convolución. Micelánea Matemática **24**, 1–14 (1996)
5. C.A. Monje, Y. Chen, B.M. Vinagre, D. Xue, V. Feliu, *Fractional-Order Systems and Controls* (Springer, Berlin, 2010)
6. A.A. Sonin, *The Physical Basis of Dimensional Analysis* (Department of Mechanical Engineering MIT, 2001)
7. G.W.S. Blair, *Elementary Rheology* (Academic Press, 1969)